EXAM QUESTIONS & ANSWERS

KAPLAN AEC
ARCHITECTURE

This publication is designed to provide accurate and authoritative information in regard to the subject matter covered. It is sold with the understanding that the publisher is not engaged in rendering legal, accounting, or other professional service. If legal advice or other expert assistance is required, the services of a competent professional person should be sought.

President: Roy Lipner
Vice-President of Product Development and Publishing: Evan M. Butterfield
Editorial Project Manager: Michael J. Scafuri
Production Coordinator: Daniel Frey
Creative Director: Lucy Jenkins

© 1980 by Architectural License Seminars, Inc.
© 2004 by Dearborn Financial Publishing, Inc.®

Published by Kaplan AEC Architecture
a division of Dearborn Financial Publishing, Inc.®
A Kaplan Professional Company
30 South Wacker Drive
Chicago, IL 60606-7481
(312) 836-4400
http://www.ALSonline.com

All rights reserved. The text of this publication, or any part thereof, may not be reproduced in any manner whatsoever without permission in writing from the publisher.

Printed in the United States of America.

04 05 06   10 9 8 7 6 5 4 3 2

CONTENTS

Introduction v

The Exams vii

Exam Questions ix

Acoustics 1

Architectural History 5

Architectural Programming 17

Building Code 23

City & Site Planning 27

Concrete 33

Construction Situations 37

Design Theory 43

Documents 47

Doors, Windows, & Glass 55

Earthquake 59

Electricity & Lighting 63

Energy Conservation 67

Financing 71

Finishes 75

Handicapped 79

HVAC 83

Legal & Zoning 87

Masonry 93

Metals 95

Moisture Protection 99

Parking 103

Plumbing 107

Sociology 111

Solar Energy 113

Structures 117

Vertical Transportation 129

Wood 133

INTRODUCTION

Welcome to Kaplan Architecture! Kaplan recently acquired Architectural License Seminars (ALS), the oldest and most respected provider of Architect Registration Examination (ARE) study products. Over 50 percent of the registered architects in practice have used ALS material to prepare for their exam. Kaplan, Inc. is the nation's leading provider of lifelong education, having served more than 30 million individuals over the past 60 years.

THE ARE

Kaplan AEC Architecture provides the only complete, centralized source for all nine divisions of the ARE. All 50 states, 5 territories, and participating Canadian provinces offer the uniform NCARB Architect Registration Examination (ARE). This exam consists of nine divisions:

- Pre-Design
- General Structures
- Lateral Forces
- Mechanical & Electrical Systems
- Building Design/Materials & Methods
- Construction Documents & Services
- Site Planning
- Building Planning
- Building Technology

The Site Planning, Building Planning, and Building Technology divisions require graphic solutions to vignette graphic problems, while the other six divisions consist of multiple-choice questions. The exams are all administered by computer. Candidates must pass all divisions of the ARE in order to become a registered architect. Those who do not pass a division of the exam may retake it after six months. For further details on the ARE, please visit the NCARB Web site at www.ncarb.org.

Kaplan AEC Architecture provides a variety of study material to help you prepare for the exam: self study courses, computer mock exams, paper mock exams, question & answer handbooks, video workshops, and a CD-ROM-based flash card system. All of our products can be ordered online at www.ALSonline.com, or by calling (800)420-1429.

The ARE is not an easy exam! Although we cannot guarantee a passing grade, we **can** guarantee our material will better prepare you for the ARE.

Good luck on your examination, and in your professional career.

THE EXAMS

Over the years, we have found that one of the most important learning tools in preparing for the architectural licensing exam consists of simulated examination questions. *Exam Questions & Answers* is designed to help candidates prepare for the Architect Registration Examination by presenting the most comprehensive collection of simulated examination questions and detailed answers that we have ever published in one volume. These questions, in the form of the actual exam, cover every aspect of architecture and are arranged alphabetically by subject, for convenient use.

The number of questions per subject is roughly in proportion to that subject's importance on the exam. For example, two divisions of the ARE are devoted to Structures. Thus, we have provided 25 Structures questions in this book. More specialized subjects that comprise a narrower spectrum of exam questions, such as Parking or Sociology, are allotted fewer questions in this volume. One more thing: the subject categories in *Exam Questions & Answers* do not match the names of the various parts of the licensing exam. The reason is that regardless of the exam's title or the examiners' good intentions, questions on a given subject are liable to show up anywhere. For example, there may be Plumbing questions in the Mechanical and Electrical Systems division or the Building Design/Materials and Methods division. So, grouping questions by actual subject appears to be the most logical arrangement.

We do not suggest that studying the questions in this book will guarantee a passing grade on the licensing exam. However, a review of this volume will add to your growing fund of knowledge, and will also indicate those areas in which further study may be worthwhile. Using this book in conjunction with other, more detailed study material will produce the greatest benefit. Candidates are encouraged, therefore, to use this book as an important reference, but they should be aware that a great deal of disciplined study is also required. Those who pass these exams are those who are well prepared.

As with our self study courses, all the material in this book has been prepared by the staff of ALS and is intended as a supplement to those courses.

We wish you success on the licensing exam and hope we have made your preparation for it a bit easier.

EXAM QUESTIONS

Except for Building Planning, Building Technology, and Site Planning, all of the exam divisions consist of multiple-choice questions. Each multiple-choice question comprises a statement followed by four choices, one of which is assumed to be the correct answer. The variety of question types possible within the multiple-choice format is remarkably diverse and may include negatively-worded questions, ranking-type questions, combination-answer questions, matching questions, and more.

In preparing for these examinations, candidates should devote some time to analyzing the multiple-choice format, since familiarity with various question types will save valuable time at the actual exam and enhance one's chances of passing. The strategy in answering any multiple-choice question consists of several progressive steps in which the candidate must:

- **read** the entire statement of the question,
- **understand** what one is seeking,
- **read** all four answer choices,
- **eliminate** the probable incorrect answers, and
- **choose** the best answer among those that remain.

If you attempt to eliminate any step in this process, you may very likely end up with the wrong answer. For example, during the actual exam, candidates may spot a key word in the statement of the question, and without reading the entire statement, quickly associate it with one of the answers—often the wrong one.

Consider the following question:

Accelerators comprise a common type of concrete admixture. Among the following statements concerning these substances, which is NOT true?

A. *One of the most common accelerating admixtures is calcium chloride.*
B. *One of the least expensive accelerating admixtures is entrained air.*
C. *Accelerators are commonly used to increase the rate of early-strength development.*
D. *Accelerators are generally employed more during winter concreting than during the summer.*

As you can see, the question is about accelerators, and in statement A the prominent term, calcium chloride, jumps out as the most popular of all accelerators. Under the stress of the exam, you might make this quick association and select statement A as the correct answer.
However, if you remember that we are looking for the *untrue* statement, you will select the correct answer B. In this case, you must eliminate all of the true statements to arrive at the one false statement, which is the correct answer.

Referring to the same question, let us suggest an expedient approach if your knowledge of accelerators is relatively meager. You might reason in the following way: Choices A and B are a total mystery; choice C, however, sounds right. Accelerate means to speed up, and if the concrete setting process were speeded up, the rate of early-strength development would probably be increased. If one wanted to speed up the concrete setting process, it would likely happen during cold, rather than hot weather, since concrete sets more slowly in cold weather. Therefore, choice D also sounds like a correct statement. We are still left with choices A and B, but by employing a minimum of technical knowledge, we have increased the chance of selecting the right answer from 25 percent to 50 percent—a definite improvement.

In using this book, we suggest that you attempt to answer each question before referring to the explanation that follows. In this way, you will be able to determine your own grasp of the subject while developing the thought process necessary to reason out the right answer. This will also help you to become familiar with the form of questions found on the actual exam and to acquire a *feel* for the actual testing experience. Having the knowledge is essential, but understanding how the exam works is no less important to a successful performance.

ACOUSTICS

1. Reverberation time depends largely on the sound absorption properties of a room. In an auditorium, the sound absorption of the audience will generally affect this time
   A. to a small, but measurable degree.
   B. slightly more than the seats they occupy.
   C. to a varying degree, depending on the type of clothing they are wearing.
   D. more significantly than the entire empty auditorium space.

Reverberation is the sound that continues in a room after its source has been cut off. It is an accumulation of overlapping echoes that produce a continuous sound. Reverberation time is the amount of time it takes a sound to decrease from a standard intensity to the point at which it becomes inaudible. Control of reverberation time in an auditorium is of paramount importance, since excessive reverberation garbles speech, distorts music, and generally creates an acoustic nightmare for performers. Control of reverberation time is achieved by using sound-absorbing materials, such as carpeting, fabrics, and acoustical tiles. Another important source of sound absorption in an auditorium is the audience itself, which accounts for a considerable part of the total absorption. In general, and regardless of their clothing, an audience affects the reverberation time far more significantly than the entire empty auditorium (correct answer D).

2. A loud office has a noise level intensity of 70 decibels. In order to reduce the noise by 50 percent, as perceived by the average ear, the noise level would have to be reduced by
   A. 8 decibels.
   B. 17 decibels.
   C. 35 decibels.
   D. 70 decibels.

The loudness of sound depends not only on intensity, but also on the frequency of the sound and characteristics of the human ear. The range of intensity to which the ear responds is enormous; a very loud sound can develop two and one half trillion times the intensity of a barely audible sound. Sound intensity is measured in decibels, and the ratio between two sounds has an approximate logarithmic relation. Because of this logarithmic relationship, 40 decibels is not twice as loud as 20 decibels; it is actually more like 100 times as loud. As far as the office in our question is concerned, an eight decibel reduction—from 70 db to 62 db—is approximately a 50 percent reduction in the noise level (correct answer A).

3. The most effective way to control noise transmission through a partition is to
   A. attach the wall covering with resilient mountings.
   B. provide an air space between dual partitions.
   C. construct the partition with very heavy materials.
   D. cover the surfaces of the partition with sound-absorbing materials.

When sound originating in a room strikes all of the room's surfaces, part of it is reflected from the surface, part is absorbed by the surface, and part is transmitted through the surface. The reduction in sound energy of the part that is transmitted is called the transmission loss, which is a measure of a partition's (or ceiling's or floor's) effectiveness in insulating a room against the transmission of neighboring sounds. In each of the choices above, the solutions stated will reduce transmission and improve insulation efficiency. However, significant sound transmission loss can only be achieved with mass, or increased weight of the partition (correct answer C). Weight of the partition per unit of area is the most important factor in determining its sound insulation efficiency, and the kind of material or the way it is held in position are of secondary importance.

4. With regard to effective acoustical design, select the incorrect statement from among those which follow.
    A. Spaces that are designed as pure geometrical shapes should be avoided.
    B. Concave wall surfaces within spaces are preferable to convex wall surfaces.
    C. Reverberation is rarely a problem in a very small space.
    D. In very large spaces, reverberation time can be as long as a half minute.

Effective acoustical design considers the total problem, not just cosmetic surface treatments. Beyond acoustical tile on the ceiling or carpeting on the floor, good design involves the type of construction to control sound transmission, as well as the shape of a space to control reverberation. Perfect geometrical shapes, such as circular plans or cubic spaces for example, are configurations that can lead to acoustical "hot spots," or repeated reflections of sound. This sort of reverberation is almost always objectionable and, therefore, should be avoided (A). The sound that strikes concave wall surfaces is also difficult to control; however, convex surfaces are highly desirable, since they diffuse sound very effectively (incorrect statement and correct answer B). Regarding statements C and D, reverberation is greatly affected by the size of a space. When the time interval between a sound and its reflection is too short, a true echo cannot be distinguished, and that is why no echoes are possible in a very small room. On the other hand, in very large spaces, such as some cathedrals and train stations, the reverberation time has been found to be as long as 30 seconds.

5. The president of a large company has complained that normal conversations emanating from an adjacent office can be overheard in his office. You have been retained to correct the situation, and consequently you recommend several possible solutions. Which of the following suggestions might be included in your recommendation?
    I. Alter the STC of the separating wall
    II. Apply a fabric covering on the separating wall
    III. Increase the number of total absorption units in the room
    IV. Provide recorded background music in the room

    A. I only   C. II, III, IV
    B. I, III, IV   D. I, II, III, IV

Speech privacy is an important concern in offices. Conversations overheard from the other side of a partition are not only annoying, they also cause concern in the mind of the listener that his own conversations may be overheard. When situations occur, such as that described in this question, the designer has several effective tools to solve the problem. First of all, the sound transmission class (STC) of the separating wall can be increased, generally by increasing the density of the wall (I). Applying a fabric covering on the wall (II) will do little to prevent sound transmission. At best, this application will reduce the reverberation time of the room by absorbing a small part of the original sound. Increasing the absorption units in the room (III) will also be effective, especially if this includes the addition of heavy carpeting and drapery. Finally, the introduction of background music will tend to mask the unwanted sound and render it unintelligible. This kind of music should include a mixture of all frequencies in the audible range. The correct combination of choices is included in correct answer B.

6. In designing a concert hall the acoustical engineer would, most likely, recommend a reverberation time of approximately

   A. zero.
   B. two seconds.
   C. five seconds.
   D. ten seconds.

Sound travels at a speed of about 1,130 feet per second. If the walls and ceiling of a room are not absorbent, that is to say, if they are not covered with sound-absorbing draperies or porous materials, the sound will not just reflect once from each wall, but will bounce from wall to wall, passing by the listener's ear many times in a second and creating many modes of vibration in the room. Obviously, the sound loses some energy at the walls and while traveling through the air, and thus, the level of the sound decreases each time it is reflected and crosses the room. In acoustics, this is known as reverberation, and reverberation time is defined as the number of seconds it takes for the sound to fade below hearing level. Ideally, the reverberation time of a concert hall should be adjusted to suit the style of music being performed. Most modern halls, however, serve a wide variety of music, and the architect must therefore seek the best compromise. Acceptable median reverberation times range from 1.5 seconds to 2.0 seconds (correct answer B). Shorter reverberation times produce a dry and lifeless sound, whereas longer periods produce overlapping sounds that lack clarity and definition and sound blurred or muddy. A reverberation time of zero, choice A, is unrealistic and generally unattainable; five to ten seconds, choices C and D, is extremely long, and results in very poor acoustics.

7. Which of the following statements in regard to NC curves is correct?

   A. Low frequency sounds are permitted to have a greater sound intensity than high frequency sounds.
   B. The permitted sound intensity is the same for sounds of all frequencies.
   C. High frequency sounds are permitted to have a greater sound intensity than low frequency sounds.
   D. The NC number varies inversely as the sound pressure level.

NC curves are widely used in specifications as criteria for the maximum noise level in a given space under a given set of conditions. Each curve represents the maximum sound pressure level in decibels for any given

frequency. Greater sound pressure levels are permitted at low frequencies than at high frequencies (correct statement A). For example, an NC 30 curve indicates a maximum sound pressure level of about 41 decibels at a frequency of 250 Hertz, but this level drops to about 27 decibels when the frequency increases to 8,000 Hertz. As the NC number increases, the maximum sound pressure level also increases (incorrect statement D). Thus, an NC 20 curve might be specified for a quiet space, such as a broadcast studio, while an NC 40 curve would be more appropriate for a noisier space, such as a retail store.

ARCHITECTURAL HISTORY

1. The carved likenesses of great American presidents on Mount Rushmore in South Dakota have precedents that go back in history over 3,000 years. Foremost among the early archetypes was

   A. the Great Temple at Abu Simbel.
   B. the Temple of Ammon at Karnak.
   C. the Temple of Horus at Edfu.
   D. the Theater of Dionysus at Athens.

At Mount Rushmore, in the Black Hills of South Dakota, the heads of Washington, Jefferson, Lincoln, and Theodore Roosevelt are sculptured directly on the mountain's granite face. Their enormous scale, proportional to men nearly 470 feet tall, makes them visible for several miles. About 1300 B.C., the Pharaoh Rameses II created an equally prodigious appearance for his Great Temple at Abu Simbel (correct answer A). The temple facade consisted of four colossal seated statues of Rameses, each 65 feet high, which were sculptured out of the living rock of the mountain. This incredible work of art was recently removed to higher ground as a result of the Aswan Dam project. The Temples of Ammon and Horus (B and C) were freestanding Egyptian structures with no large sculptures, and the Greek Theater in Athens (D), although carved out of the sloping mountainside, contained no sculptured pieces at all.

2. Which of the following statements regarding the Bauhaus is false?

   A. The principal purpose for establishing the Bauhaus was to effect a merger between art and technology, which would serve as a basis for modern design.
   B. The Bauhaus movement was directed in Dessau by Walter Gropius, in Berlin by Mies van der Rohe, and later in Chicago by Lazslo Maholy-Nagy.
   C. The Bauhaus philosophy advocated both artistry and craftsmanship, the use of machinery to produce well-designed objects, and a collective effort during creation and production.
   D. Although the Bauhaus had tremendous early influence on modern design theory, its sudden dissolution in 1933 led to the immediate decline of its prestige and usefulness in the architectural world.

How much one must know about a significant historical movement is clearly revealed by a question such as this one. Most candidates probably know that the Bauhaus was a prominent school of design founded by Walter Gropius in Germany, in 1919. That information is not enough, however, to answer this question. One must also know something about its purpose, its philosophy, and its short but influential history. The Bauhaus purpose and philosophy are accurately expressed in choices A and C. Choice B correctly describes the sequence of directors who headed the Bauhaus, first in Germany and later in Chicago, where the New Bauhaus was re-established following its termination by the Nazis in 1933. The one false statement we are looking for, therefore, is correct answer D, which suggests that dissolution of the Bauhaus led immediately to the decline of its worldwide influence. Quite to the contrary, after the school in Germany was closed, its teaching and methods continued to exercise an enormous influence which endured for a great many years. Both in Europe and the United States, scores of schools adopted the Bauhaus teaching methods; and throughout the world, the Bauhaus ideas and products had a tremendous impact.*

3. The traditional Early Christian basilican churches were generally distinguished by
   A. regular pier-buttressed walls.
   B. pointed arched openings.
   C. timber trussed roofs.
   D. barrel vaulted naves.

   Candidates are required to know some of the salient features which characterized each of the traditional architectural styles. In this question, for example, one should be aware that buttresses (A) and pointed arches (B) did not generally appear in church design until the Medieval period. Vaulted naves (D), as well, were developed somewhat later in ecclesiastical history. It was the timber trussed roofs, therefore, (correct answer C) which were almost universally applied to Early Christian basilican churches.

4. The World's Columbian Exposition of 1893 in Chicago was extraordinary for the effect it had on American architecture. Among other things, it put an end to the remarkable strides made by the Chicago School, it signaled the end of Sullivan's architectural career, and it served as the major influence in almost all public buildings for years afterwards. The architectural style of this influential Exposition was
   A. Modernized Roman.
   B. Renaissance Revival.
   C. Eclectic Classic.
   D. Neo-Baroque.

   The chief architect of the 1893 Chicago World's Fair was Daniel Burnham, a leader of the progressive Chicago School. The Fair's design, however, was dictated by New Yorker Charles McKim, who was a strong proponent of Roman work. The composition of the Fair was actually Neo-Baroque (correct answer D), with crossed axes, rigid formality, and facades that derived from the Beaux-Arts school in Paris. The style was called "Modernized Classic," but in fact it was not modern, it certainly was not classic, nor did it emanate from Rome or the Renaissance. It was a fairly original eclectic expression that recalled the Italian Baroque more than any other formalized style.

5. Match the following architectural terms with the historical styles with which they are commonly associated.

   I.   Stoa   1. Medieval
   II. Dolmen   2. Greek
   III. Tuscan   3. Modern
   IV. Brise-soleil   4. Roman
   V.   Machicolation 5. Prehistoric

   A. I-4, II-1, III-2, IV-3, V-5
   B. I-2, II-5, III-4, IV-3, V-1
   C. I-4, II-3, III-5, IV-2, V-1
   D. I-1, II-5, III-4, IV-2, V-3

It is inevitable that a few architectural history questions will require candidates to know the meaning of some relatively obscure historical terms, or as in this question, at least the period with which they are associated. Candidates are advised, therefore, to review a glossary of terms as part of their overall study pattern. In this question, one should know the meaning of at least a couple of the terms in order to discover the answer. Perhaps the most obvious term here is Tuscan (III), which was a simplified architectural order developed by the Romans (4). Knowing this fact alone, one could eliminate choices A and C, in which the term Tuscan is improperly matched. Possibly one knows that Brise-soleil (IV) refers to the permanent grid of sun shields covering the windows of a building, which was first used by Le Corbusier and Oscar Niemeyer in the Ministry of Health building in Rio de Janeiro in 1937 (3). This additional fact would eliminate choice D, and thus, one would be left with choice B, which is the correct answer. Knowledge of any of the other facts would, of course, verify the correctness of this choice. For example, Stoa (I) is a detached portico in Greek architecture (2), Dolmen (II) refers to the prehistoric arrangement (5) of two or more upright stones supporting a slab, and finally, Machicolation (V) refers to the projecting parapets of Medieval structures (1) that had floor openings through which molten material was dropped on the enemy below.

6. The vast majority of English cathedrals were constructed during the Gothic period. One of the following cathedrals, however, was built in another age and in another style. Which one was it?

   A. Salisbury Cathedral
   B. Canterbury Cathedral
   C. York Cathedral
   D. St. Paul's Cathedral, London

It is unfortunate that many questions on this exam require the recollection of a relatively trivial fact. This type of question teaches one very little; it also ignores the valuable lessons to be found among the rich examples which make up our architectural heritage. Nevertheless, we include such a question here to alert candidates to this situation. The correct answer is D. It was Christopher Wren's masterpiece, St. Paul's, that was built during the Renaissance period to replace the Medieval cathedral destroyed in London's great fire of 1666.

7. Daniel Burnham, Baron Haussmann, and Le Corbusier all had in common which of the following?

   A. Each was trained in the academic tradition of the Beaux-Arts in Paris.
   B. Each had a profound effect on the theory and practice of modern city planning.
   C. They were all designers of influential planned residential developments.
   D. They were all leaders in the industrial-technological revolution in architecture.

The three designers, whose lives spanned three different eras, all made a significant contribution to the art of city planning (correct answer B). At the turn of the century, during the City Beautiful era, Burnham created city plans for Chicago and elsewhere which were based on French Renaissance principles. Regarded as the father of American city planning, Burnham's philosophy was embodied in his dictum, "Make no little plans..." During the mid-19th century, under Napoleon III, Haussmann was charged with the rebuilding of Paris. He created the tree-lined, broad boulevards that determined the present face of the city, and which were subsequently much copied. From the 1920s on, Le Corbusier, through his plans and writing ("La Ville Radieuse") exerted an enormous influence on town planning. His proposals involved widely-spaced high-rise structures, which were separated by large open green areas.

8. Many of the enormous Roman structures would have been impossible to construct without the use of groined vaults. This ingenious development accomplished all of the following, with the exception of

   A. it concentrated the thrust of the vaults at the four corners of the vaulting bay.
   B. it made possible immense arches which were virtually unlimited in their span.
   C. it enabled large windows to be inserted high up under the arches of the vaults.
   D. it enabled the timber centering used for one bay of the vaulting to be dismantled and used again.

The Roman groined vault was ingenious indeed. It consisted of two intersecting semi-circular vaults (tunnel or barrel vaults), which concentrated the vault thrust at the corners (A). By doing so, the sides could be opened, which allowed windows to be inserted (C), or other square vaulting bays to be joined together in a continuous rectangular space. It was also an economical construction system, as temporary timber forms (centering) could be re-used almost endlessly (D). As wonderful a development as groined vaults were, however, it did not permit unlimited spans (correct answer B). Roman stone arches, although often reaching spans of over 80 feet, were limited by the compressive strength of the stone, which could crush when supporting an excessive amount of superimposed stonework.

9. Architectural design theory was significantly influenced in the mid-1960s by the controversial book, *Complexities and Contradictions in Architecture*. In this book Robert Venturi said, "I am for messy vitality over obvious unity; richness of meaning rather than clarity of meaning." Venturi was implying that

   A. architecture should not take itself too seriously.
   B. architecture frequently involves compromise and the principles of accommodation.
   C. rigid and unified architecture lacks vitality and meaning.
   D. rational modern architecture is dead.

Questions such as this can create nightmares for candidates, especially if they are unfamiliar with the author's philosophy, since quotations are generally taken out of the context of a larger work. In this case it would certainly help to know that Venturi believed in the principles of accommodation and in the acceptance, if not actual admiration, of the ordinary, pluralistic world in which we all live. In all this, he did not see the "mess" or the inconsistencies, but rather he observed the richness and vitality. Thus, architecture must often be accommodating and compromising (correct answer B). This does not mean, however, that architecture should not be a serious business (A), nor that a more rigid expression cannot also have significance and vitality (C). Least of all did Venturi imply that modern architecture was dead; that was a conclusion of certain post-modern critics, who believed that Venturi's influence had forever changed the direction of contemporary architecture.

10. The "Chicago window" might be best described as an opening having

    A. a fixed center and movable sides.
    B. a large, unbroken expanse of glass.
    C. a vertical proportion and round arch top.
    D. a horizontal proportion and many small panes.

The desire to provide an abundance of natural light for commercial office buildings gave architects the incentive to reduce the structure as much as possible and fill the voids of the slender structural cage with glass. The resulting "Chicago window" became a large horizontal expanse, with fixed center and movable sides (correct answer A), which stretched the entire length of a structural bay. This detail became one of the distinguishing characteristics of the Chicago School projects during the 1880s and 90s.

11. Match the following historical styles with the type of masonry expression commonly associated with each style.

    I. Early Greek
    II. Prehistoric
    III. Renaissance
    IV. Roman
    1. megalithic masonry
    2. decorative masonry
    3. cyclopean masonry
    4. rusticated masonry

    A. I-2, II-1, III-4, IV-3
    B. I-1, II-4, III-3, IV-2
    C. I-3, II-1, III-4, IV-2
    D. I-3, II-1, III-2, IV-4

Examiners are often able to test several architectural facts in a single question, and in those cases, guessing the answer can be especially difficult. In this question, for example, one must be able to associate at least three kinds of masonry with their appropriate historical styles. The analysis of this type of question can begin almost anywhere; however, you may notice that in the four answers there are only two choices for (II) Prehistoric—it is matched with either (1) megalithic or (4) rusticated. If you can remember that megalithic refers to the great (mega) stone (lith) structures that were constructed during prehistoric times, then you can eliminate answer B, which matches rusticated with Prehistoric. Perhaps you recall that rusticated masonry was a Renaissance method used to arrange rough-surfaced stones. In that case we can eliminate answer D; and now we are left with only two choices, A and C. The difference between answers A and C lies in whether decorative or cyclopean masonry belongs with Greek or Roman architecture, and knowledge of either fact will result in the correct answer. Cyclopean masonry refers to large, rough stones piled one upon another, and examples of this work were found at both Knossos and Mycenae in ancient Greece. The last item, decorative masonry is a fairly imprecise term, since most masonry throughout architectural history has been decorative to some extent. However, the ancient Romans enjoyed the richness of architectural decoration as no other civilization until the Baroque period. As Fletcher stated, "The Romans never seem to have been satisfied till they had loaded their monumental buildings with every possible ornamental addition." The correct match, therefore, is found in answer C.

12. Which of the following statements concerning Art Nouveau is false?
   A. Art Nouveau originated in Paris, where it reached its most brilliant heights.
   B. The style expressed an essentially decorative trend, which emphasized the ornamental value of curvilinear lines and undulating forms found in nature.
   C. The movement arose out of a determination to create a new style, which would resist the rising influence of industrialism and counteract the new technology of the machine age.
   D. The romantic Art Nouveau movement lasted a mere 20 years, but its doctrine affected the artistic expression of the entire European continent.

The Art Nouveau movement is accurately described in statements B, C, and D. Statement A is false, and therefore the correct answer to this question. The style did, of course, reach Paris, where many fine examples remain to this day; however, it had its origins in Belgium, not France. Although the Art Nouveau did help terminate the great age of eclecticism, the emphasis of the movement was towards the romantic world of craftsmanship and away from the machine, and hence, also away from truly modern architecture.

13. Designed on a circular module, the structure achieved the liberation of space the architect had been searching for. Mushroom-shaped columns rose up to become a part of the ceiling, which produced a feeling of spatial freedom with only the slightest sense of enclosure. It has been called a splendid achievement; noble, Roman, and streamlined, all at once. The building described is the

A. Pantheon, Rome
B. Treasury of Atreus, Mycenae
C. Johnson Wax Building, Wisconsin
D. Temple of Vesta, Rome

There are some exam questions that might be considered deceptive, and this is an example of such a question. If one were to take the building description literally and focus on the words "circular" and "Roman", one might come up with either the Pantheon (A) or the Temple of Vesta (D) as the answer, since both Roman structures were circular in plan. On the other hand, the Treasury of Atreus (B) was also circular in plan; and moreover, mushroom-shaped columns were a product of this Mycenaean period of early Greek architecture. However, none of these ancient structures fits the description exactly, as does the Johnson Wax Building by Frank Lloyd Wright. In this structure, built in 1936, Wright created one of the masterful spatial experiences of his career. It was lighted through glass tubes from above, as in a Roman building; and as the columns became part of the ceiling, continuity was its distinguishing quality.

14. Between the 16th and 19th centuries in Europe, a great variety of distinct architectural styles emerged. Listed below are four of these styles followed by a single descriptive word which was characteristic of each. Which one is least accurate?

A. Mannerism—contrived
B. Historicism—eclectic
C. Neo-classicism—irrational
D. Romanticism—nostalgic

A peculiarity of historians is that they produce endless labels to identify each different architectural expression. The problem for architectural candidates is that they must know what all these labels mean. Here one must identify the only descriptive word which does not apply to the style with which it is joined; and for that, one must know something about each of the styles. Mannerism (A) was the 16th century Renaissance development that produced arbitrary, manipulated, and frequently illogical compositions. In every sense of the word, it was contrived. Historicism (B) describes the activities during the first half of the 19th century, when a great variety of styles were employed. It was the great age of Eclecticism, which embodied styles borrowed from all past historical periods. Romanticism (D) emerged at the beginning of the 19th century as an expression of discontent among all artists. It was an escape from the realities of the Industrial Revolution, and it involved all the conventional romantic qualities, among which nostalgia was paramount. The answer to the question, by elimination therefore, is C. The neo-classicism of the 18th century was not irrational, as was the Baroque style it supplanted; rather, it was an expression of architectural restraint, logic, and—above all—rationalism.

15. Aside from the Egyptian period, which extended over 3,000 years, the longest continuous period of architectural development occurred during the

   A. Roman Empire.
   B. Renaissance Period.
   C. Gothic Period.
   D. Byzantine Empire.

The Byzantine Empire (correct answer D) persisted for over a thousand years, and thus, was the longest period of architectural development after Egypt. The Roman emperor Constantine moved the capital from Rome to the old Hellenic town of Byzantium in the year 334. At that strategic location, the great Christian and Latin capital was developed. The marriage of Roman and Eastern forms created one of the greatest architectures of history. The Byzantine capital endured from the 4th to the mid-15th century, at which time it fell to the Turks.

16. Throughout the middle of the 19th century in America, literally hundreds of Gothic Revival churches were built in small towns throughout the country. To one degree or another, they were inspired by the work of

   A. Thomas Jefferson.
   B. Richard Upjohn.
   C. Henry H. Richardson.
   D. Charles Bulfinch.

It was Upjohn (correct answer B), famous architect and champion of the early American Gothic Revival, whose great influence extended, literally, from sea to sea. His most famous work was Trinity Church in New York, built around 1840, and patterned after an English country parish church. In this and subsequent churches, most of which were more modest, as well as more rustic, Upjohn's work helped spread the message of Medieval art. Jefferson (A) was a strong proponent of Roman architecture; Richardson (C), who was architect of another Trinity Church, in Boston, was inspired by French Romanesque; and finally, Bulfinch (D) was the country's leading Greek revivalist.

17. For a number of logical reasons, the earliest Christian baptisteries were housed in

   A. former Roman tombs.
   B. former Roman baths.
   C. centralized structures.
   D. basilican structures.

Baptisteries were originated for the sacrament of baptism, which was a condition of one's entry into the Christian Church, as well as a precondition of resurrection after death. Because of the powerful theological association with burial buildings which were round or square, a centralized structure (correct answer C) was invariably used. The association with bathing or immersion also led early Christians to imitate the Roman thermae, whose principal rooms were also centralized. Finally, from a standpoint of pure function, a round or octagonal structure best served the needs of the baptismal ceremony.

18. Which of the following statements concerning the use of iron in architecture are true?

I. The first use of prefabricated iron framing in architecture was found in Paxton's Crystal Palace.

II. One of the leading proponents of the use of iron in construction was the Englishman John Ruskin.

III. Henri Labrouste, in his Paris libraries, created the earliest interiors employing the aesthetic of metal construction.

IV. Although the earliest architecture using iron framing evolved in England, the first metal-framed tall buildings were actually developed in Chicago.

V. A widespread 19th century idea held that iron used in construction was completely incompatible with "architecture."

A. I, IV
B. II, III, V
C. III, IV, V
D. I, III, IV, V

During the mid-19th century, the use of iron created a controversy about what constituted real architecture. The general belief was that iron was acceptable for bridges and industrial buildings, but it certainly had no place in "architecture" (V). Two achievements which helped transform public opinion were Labrouste's libraries in Paris (III), and the remarkable Crystal Palace in London (I). By the latter part of the 19th century, when metal-framed tall buildings were developed in Chicago (IV), the material had become more or less architecturally acceptable. Among the violent opponents to the use of iron in construction was John Ruskin, who wrote in "Stones of Venice" that architecture was something quite different from a "rat hole or railway station." Statement III, therefore, is the only false statement above, and the true statements are found in correct answer D.

19. Which of the classic columns listed below is identified by the following description? The column was sturdy and dignified, with a plain, square abacus at the capital; a fluted column shaft which exhibited entasis; and lacking a base, it rested on a stylobate.

A. Greek Doric
B. Roman Doric
C. Roman Tuscan
D. Greek Ionic

It would be difficult to guess the answer to this question if one had no idea what distinguished the column of one classic order from another. Perhaps the key descriptive word here is "sturdy," since the Greek Doric column (correct answer A) had the stumpiest proportion (diameter to height) of all the classic orders. The Roman Doric (B) was similar to the Greek, but the Romans added a base to the column and occasionally used it unfluted. The Roman Tuscan (C) also had a base and was always unfluted, while the Greek Ionic (D) was distinguished by the volutes or scrolls used on the capital. Incidentally, candidates should know that entasis refers to the outward curve of the column shaft, and that stylobate means stepped base.

20. Which of the following individuals produced major works of art that materially affected the public buildings of which they were a part?

I. Daniel French
II. Lee Lawrie
III. Pablo Picasso
IV. Alexander Calder

A. I, II
B. II, IV
C. II, III, IV
D. I, II, III, IV

Candidates should not be surprised to discover that they must know something about developments in related fields, especially as these developments have influenced architectural expression. In this question, for example, one must be familiar with four different artists, all of whom produced skillful sculptured works (correct answer D), which virtually transformed the buildings with which they were integrated. Daniel French (I) was the sculptor of the Lincoln statue housed in the famous Lincoln Memorial in Washington, designed by Henry Bacon. Lee Lawrie (II) was the sculptor of the decorative figures on Bertram Goodhue's State Capitol at Lincoln, Nebraska. Picasso (III) created a unique sculpture, which has become a tourist attraction in itself, in the court of Chicago's Miesian Civic Center. Finally, Calder (IV) produced one of his most exciting mobiles, dominating the East Wing of the National Gallery in Washington D.C., designed by I.M. Pei.

21. Select the correct statement.
    A. Gothic architecture originated in France and was primarily ecclesiastical.
    B. Renaissance architecture originated in Italy and was primarily royal and mercantile.
    C. Neither of the above statements is true.
    D. Both of the above statements are true.

Both statements are obvious oversimplifications, but essentially correct, as is answer D. One should be aware that all such simplistic statements can be misleading, since a great many Gothic castles, as well as churches, and numerous Renaissance churches, as well as palaces, were designed and built.

22. The edict "less is more" has been ascribed to Mies van der Rohe, but actually it expressed ideas that were developed throughout Europe during the first decade of the 20th century. Its closest meaning implied that one should
    A. eliminate ornamentation from useful objects.
    B. eliminate the irrelevant and simplify the essential.
    C. concentrate on space and mass, the two essentials.
    D. pursue a more essential architecture.

This question is difficult to answer because there is truth in each of the choices, although Mies' quotation was explained by his own words in correct answer B. It was the Austrian Adolf Loos who wrote about the crime of ornamentation (A), and a similarly limited view was expressed by the Dutch architect H. P. Berlage, in choice C. The general advice in choice D came from Otto Wagner, a Viennese architect who pursued his "more essential" architecture by eliminating Renaissance trappings from his buildings. All of these men were reacting against the decorative excesses of the Victorian period. It was Mies, however, whose singular vision, ability, and consistency fashioned a rich aesthetic expression by relying on the quintessence of architecture.

23. Which of the following modern tall buildings is mismatched with the special feature associated with it?
    A. Ford Foundation Building—full height interior garden
    B. John Hancock Building (Chicago)—twin towers in a park-like setting
    C. Lever House—open ground floor plaza
    D. Richards Medical Research Building—articulated service towers

All of the buildings are accurately matched with their distinguishing features with the exception of the John Hancock Building (correct answer B), designed by Skidmore, Owings, and Merrill. The structure is a single (not twin) battered tower that inclines inward as it rises 100 stories above Michigan Avenue. It is set in a predominantly paved shopping plaza (not a park-like setting), and the tower is further distinguished by the exposed diagonal steel braces. Candidates should be familiar with all the other structures mentioned here and their designers: Roche & Dinkeloo (A), Gordon Bunshaft—SOM (C), and Louis Kahn (D).

24. Select the correct statement concerning the Ancient Egyptian civilization.
    A. The Egyptian religion was monotheistic and guided by the Pharaoh. He, in turn, was advised by the priesthead, a group with much learning, but strictly limited authority.
    B. Monumental Egyptian buildings employed thousands of well-paid, skillful laborers who became available when the annual flooding of the Nile made agricultural work impossible.
    C. Immortality in ancient Egypt became an obsessive idea that required preservation of the body to insure survival after death.
    D. The masonry houses of the ancient Egyptians were designed to respond to the extremes in climate, which ranged from the summer's intense heat to the heavy rains and occasional frosts of the winter months.

With the exception of correct answer C, all of the statements are incorrect. To begin with, in choice A, the Egyptian religion was polytheistic, with a cult of many gods representing natural phenomena, heavenly bodies, and even animals. Furthermore, the priesthood was a very powerful group, invested not only with much learning, but with virtually unlimited authority. In choice B, workers did alternate between agricultural and construction projects, but this work force consisted principally of unskilled slave labor and prisoners of war. Finally, in choice D, the inaccuracies concern the Egyptian climate, which has been referred to as very consistent—always hot. Rain is rare, even during the winter months, and frosts are unknown.

25. The great advantage of Buckminster Fuller's geodesic dome is that it
   A. encloses the maximum volume of space while employing the minimum enclosing surface.
   B. permits total environmental control within the enclosed space.
   C. allows for flexibility of form; that is, it can be constructed as a steep or a shallow dome.
   D. provides vast quantities of universal space, which can be adapted to nearly every function.

Bucky Fuller's famous dome has become a standard part of the contemporary architectural vocabulary, largely because it is a relatively inexpensive way to create a large volume of space (correct answer A). The geodesic dome was developed by Fuller to combine the structurally desirable properties of the tetrahedron and the sphere. Thus, it is rigid in shape (C) and relatively inflexible in function (D): whether one needs or wants it, the dome always produces a circle in plan, which may not always suit the function. Concerning choice B, the geodesic dome does not permit any more efficient environmental control than any other geometric shape.

ARCHITECTURAL PROGRAMMING

1. All of the following fall within the scope of architectural programming, *except*
   A. the traffic patterns in the vicinity of the site.
   B. estimated space needs for the project.
   C. related buildings and facilities.
   D. building materials and construction methods.

Architectural programming is a process leading to the statement of an architectural problem and the requirements to be met in offering a solution. Traffic patterns (A) and space needs (B) may be considered during architectural programming, along with numerous other factors. Buildings and facilities related to the planned building by function, location, or ownership (C) might also fall within the scope of architectural programming. Building materials and construction methods (D) comprise part of the solution *to the problem, rather than the statement* of the problem, *and therefore do not fall within the scope of architectural programming. D is therefore the correct answer.*

2. During the schematic design phase, the architect determines that the budget for a project is inadequate. What is his best course of action?
   A. Reduce the quality of the building's materials and systems.
   B. Reduce the scope of the project.
   C. Notify the owner and lender and temporarily suspend work on the project until they reach a decision.
   D. Review the program with the owner to determine where cost savings should be made.

This kind of question has appeared in one form or another on a number of past exams. When faced with the situation described in this question, the architect should always review the program with the owner (correct choice D). Certain elements of the program may need to be reduced in scope or even eliminated altogether. Perhaps some reduction of quality can be made. In any event, these decisions should be made by the owner, with the advice of the architect—not by the architect acting alone (A and B), nor by the owner and lender acting without the architect (C).

3. The programmed efficiency of a building is 65 percent. If the efficiency were increased to 70 percent and the net area remained constant, the gross area would
   A. decrease by 7 percent.
   B. decrease by 5 percent.
   C. remain unchanged.
   D. increase by 5 percent.

The efficiency of a building is the ratio of the net area to the gross area, where the net area is the sum of all usable floor spaces not including circulation and general service areas. If we assign number 1 to the 65 percent efficient building and number 2 to the 70 percent efficient building, we have

$$\frac{\text{net area 1}}{\text{gross area 1}} = 0.65 \text{ and}$$
$$\frac{\text{net area 2}}{\text{gross area 2}} = 0.70$$

Since the net area remains constant,
net area 1 = 0.65 gross area 1 =
net area 2 = 0.70 gross area 2

0.65 gross area 1 = 0.70 gross area 2

$$\frac{\text{gross area 2}}{\text{gross area 1}} = \frac{0.65}{0.70} = 0.93$$

In other words, the gross area of the 70 percent efficient building is 93 percent of the gross area of the 65 percent efficient building. The gross area thus decreases by 7 percent (correct choice A).

4. Float time is
   A. the amount of time that a construction activity can be delayed without causing the project completion to be delayed.
   B. the amount of time between completion of an increment of construction and the earliest date when loads may safely be imposed on that increment.
   C. nonproductive time on a construction project, such as weekends and holidays.
   D. the shortest time required to complete a construction activity.

Critical path method (CPM) is a management tool used in the planning and scheduling of construction, in which a construction project is divided into separate identifiable jobs, called activities. These are indicated graphically on a network diagram, including the estimated time to complete each activity. Float time is the extra time available for an activity or group of activities above its estimated time duration without any resulting delay in completion of the project. A is therefore the correct answer.

5. Which of the following basic organizational concepts is matched with the statement which most closely relates to it?
   I. Single-level
   II. Multi-level
   III. Compact
   IV. Radial
   A. Economical mechanical and electrical systems (II)
   B. Protection for pedestrians from inclement weather (I)
   C. Ease of access for the physically handicapped (III)
   D. Ideal for surveillance and visual control (IV)

There are several basic organizational concepts in site and building design which are available to the designer in arranging spaces and volumes to meet the functional needs of the user. For example, if economy through the use of shortened duct and conduit runs is of primary importance, a compact arrangement of space and volume is preferred (A-III). If pedestrians must circulate under cover to and from various parts of a project, a multistory scheme will best respond to this requirement (B-II). The physically handicapped are best served if all facilities are on one level, precluding the use of vertical transportation (C-I). Visual control and surveillance, as required in libraries and nursing stations of hospitals, are best achieved in a radial scheme where the control point is located at the intersection of the radiating wings. The only correctly matched statement is found in answer D.

6. Which of the statements below does not accurately reflect the goals and objectives of the fine arts department of a major public university?

   A. To act as a catalyst for the cultural development of the community.
   B. To expose local school children to the arts.
   C. To provide a subscription concert series for students and public.
   D. To provide commercial support for promising local artists.

   Choices A, B, and C describe worthwhile objectives of a university in providing cultural enrichment for the community. However, choice D indicates that the university has some obligation in the exposition and sale of the work of local artists. Although the university may occasionally display local work, as a public institution it has neither the obligation nor the legal authority to engage in a profit-making enterprise for itself or its clients. The correct answer is D.

7. The project development budget for a proposed County Library is limited to 6.3 million dollars. The building program results in a total net usable area of 49,000 square feet with a targeted efficiency of 70 percent. Select the *incorrect* statement from those below.

   A. The total gross building area will be 70,000 square feet.
   B. The unit cost for construction will be $90 per square foot.
   C. At 8 percent the architect's fee would not exceed $500,000.
   D. The cost of furnishings and equipment would normally be included in the 6.3 million dollar budget.

This type of question tests the candidate's understanding of the various components that make up a project budget. To begin with, one must be aware of the difference between a project development budget and a construction budget. The construction budget consists of the anticipated contract price plus a small contingency allowance for change orders and other costs incurred during construction. The project budget, however, includes not only the cost of construction, but in addition, all costs for fees, surveys, tests and inspections, utility connections, furnishings and equipment, etc. Those costs over and above the basic construction cost normally amount to about 15 percent of the total project budget. The efficiency of a building is the ratio of the net area to the gross area, where the net area is the total usable floor area, not including circulation and general service. Since we are told that the programmed net area is 49,000 square feet and the efficiency is to be 70 percent, we can calculate the building's gross area as follows: 49,000 ÷ 70% = 70,000 square feet. Consequently, choice A is a correct statement. To calculate the building's unit cost, we simply divide the cost of construction by the total building area. Since we know that the construction budget is about 85 percent of the 6.3 million dollar project budget, the unit cost is not $6,300,000 ÷ 70,000 SF, but roughly 85 percent of that, or about $76.50 per square foot. Choice B is therefore the incorrect statement and the correct answer to the question. The architect's fee is usually based on a percentage of the cost of construction; we can calculate this to be 8% × .85 × 6,300,000 = $428,400 (C is correct). And, as stated before, furnishings and equipment are included within a project budget; therefore, choice D is also a correct statement.

8. Which of the following would not be considered one of the major tasks in programming a project?
   A. Gather and analyze pertinent facts.
   B. Discover and develop a concept.
   C. Develop outline specifications.
   D. Establish code requirements.

Architectural programming consists of the development of an explicit statement of an architectural problem. This statement lists the requirements to be met by a solution which the architect must develop. All of the choices except correct answer C define certain programmatic requirements. The preparation of outline specifications, however, is a response to these requirements by defining the types of materials and systems best suited to solve the problem during the schematic and design development phases.

9. A lecture hall is programmed to accommodate 300 persons in fixed auditorium-type seats. A demonstration table with sink and utilities and the adjacent lecture area requires approximately 500 square feet. Which of the following dimensions would provide an adequate space for the programmed use?
   A. 50 feet wide by 50 feet deep
   B. 40 feet wide by 65 feet deep
   C. 50 feet wide by 80 feet deep
   D. 25 feet wide by 110 feet deep

This question tests a candidate's knowledge about two basic considerations in the programming of lecture space: 1) the amount of area required to accommodate a specified number of occupants and, 2) the preferred shape of a space for a particular programmed use. Fixed auditorium-type seating requires approximately seven square feet per occupant. Allowing for the additional 500 square feet of lecture and demonstration area at the front of the room, we calculate the space need to be 300 × 7 + 500 = 2,600 square feet. The shape of a lecture hall is determined largely by sight lines, since the demonstration table must be clearly visible from all seats. Consequently, a square room is not desirable, since some of its seats will be located too far to the sides of the room, causing distorted sight lines, especially for projected images on a screen. A narrow but deep rectangular shape will provide good sight lines for half the seats, but those located in the rear portion will be too far away to see and hear well. Choice A is too small a space, choice C is too large, and choice D is a poorly proportioned space. Choice B, however, is the correct size and a good proportion for the programmed use.

10. An urban site measuring 300 feet by 500 feet has a zoning requirement for 20 foot setbacks along all property lines. Additional restrictions limit lot coverage for construction to 75 percent of the total site area and the maximum building height to 120 feet. Assuming story heights of twelve feet from floor to floor, the maximum allowable building area may not exceed
    A. 1,125,000 square feet.
    B. 1,196,000 square feet.
    C. 1,350,000 square feet.
    D. 1,435,200 square feet.

Questions requiring candidates to calculate building and site development areas based on zoning or similar restrictions often appear on the examination. In this particular case, we are given three zoning requirements and must choose the most restrictive to determine the maximum allowable building area. The first of these are the setbacks for the 300 by 500 foot site. In subtracting 20 feet times two from each side, we calculate a buildable lot area of 260 by 460 feet (119,600 SF). A further restriction is the 120 foot height limit. Based on the assumed 12 foot story dimension, we know the building is limited to ten stories. Ten times 119,600 = 1,196,000 SF of potential building area. However, another restriction limits lot coverage to 75 percent of the total site area; i.e., 500' × 300' × 75% (112,500). If, once again, we assume a 10 story building, the maximum possible building area is 112,500 × 10 = 1,125,000 square feet of building. Since the more restrictive requirements in zoning ordinances apply, choice A, the smaller allowable area, is correct.

11. The architect completed working drawings for a new hospital two years ago when the ABC Building Cost Index stood at 1,250. Because of legal problems, the project was delayed until now and the owner has instructed the architect to finally advertise for bids. The original construction budget was set at $24 million. The ABC Index now stands at 1,650. Select the *incorrect* statement from among those below.

A. The construction budget must be increased to approximately $31.5 million.

B. The architect is entitled to an adjustment in his fee for services.

C. The approximate annual rate of inflation of construction costs is 15 percent.

D. None of the above.

A national construction cost index is one way to measure the escalation of building costs on a general basis. The architect would probably prepare a revised, detailed estimate prior to going to bid and inform the owner of the required budget increase caused by the two-year delay and resultant cost increase. In choice A, we are asked to calculate the approximate cost increase based on the construction index. This is done as follows: (1,650 ÷ 1,250) × $24 million = 1.32 × 24 = $31.68 million which rounds off to $31.5 million. The assumed annual rate of construction inflation of 15 percent (C) can be checked as follows: For the first year, 1,250 × 1.15 = 1,437.50; for the second year, 1,437.50 × 1.15 = 1,653.125. Choice C is a correct answer, since 1,653.125 is very close to 1,650. Choice B is also correct, since the agreement between owner and architect, AIA Document B141 Article 6.4.1, provides for an adjustment of fee when a project is suspended for more than three months. Since all three statements are correct, choice D is the right answer.

12. Which of the following would not be considered during the programming phase of the architect's services?

A. Net to gross floor area ratios

B. Types of thermal insulation

C. Cash flow and return on invested capital

D. Off-street parking capacities

This question provides us with four factors that may be considered in the development of a project. The idea, of course, is to identify the one that is not considered during the programming phase of the project. The net to gross area ratio (A) has an effect on both the building's efficiency and its ultimate cost, both of which are of major interest to the owner. The cash flow and return on invested capital (C) are major economic considerations for any owner whose objective is to profit from his project's development. Off-street parking capacities (D) must be stated either to satisfy local requirements or owner objectives in making his facility available to potential users with cars. All of these are a part of the program and define the owner's goals and objectives. Although the acceptable amount of heat loss or gain through exterior walls and roofs may be considered during programming, insulating materials are usually selected during the design development or construction documents phases of work. The correct answer is B.

BUILDING CODE

1. The Uniform Building Code classifies every building according to which two of the following?
   I. The building's quality of construction
   II. The building's type of construction
   III. The building's occupancy
   IV. The building's proximity to Fire Department equipment
   A. I and II
   B. I and III
   C. II and III
   D. II and IV

Candidates are expected to understand the theory and application of building code regulations. Although all building codes used in this country are generally similar, they are by no means identical. The code most frequently used as the basis for questions on the licensing exams is the Uniform Building Code. According to this code, every building is classified by the building official according to its use or character of its occupancy (III) and its type of construction (II), making C the correct answer.

2. Flame-spread rating refers to
   A. a classification of finish materials based on tests.
   B. the fire-resistive rating of an area separation wall.
   C. a method of classifying fire dampers.
   D. a designation of flammable liquids according to volatility.

Chapter 42 of the Uniform Building Code permits combustible material to be used for interior finish in a building, provided its flame-spread rating, as determined by tests, complies with the code (correct answer A). There are three classes of finish materials based on flame-spread characteristics: Classes I, II, and III. Class I has the least amount of flame-spread and consequently may be used in critical areas such as enclosed vertical exitways. Class II is intermediate and is permitted in other exitways. Finally, Class III has the greatest flame-spread rating and may not be used in exits, but is generally permitted in other spaces.

3. A building houses two different occupancies. Which of the following are correct, according to the Uniform Building Code?
   I. The building is limited in size by the code requirements for the more restrictive occupancy.
   II. Occupancy separations may be vertical or horizontal or both.
   III. Every part of the building must conform to the code requirements for the more hazardous occupancy.
   IV. Occupancy separations must be of three-hour fire-resistive construction.
   A. II, IV
   B. III, IV
   C. I, III
   D. II only

When a building is used for more than one occupancy purpose, each part of the building comprising a distinct occupancy must be separated from any other occupancy as specified in the code. The occupancy separations may be vertical or horizontal or both (II). Each such portion must conform to the code requirements for the occupancy it houses. Statement II is therefore correct. I is not correct; since the occupancies are separated, the total allowable area of the building takes into account all the occupancies, not just the most restrictive. III is also false; again, because of the occupancy separation, each portion of the building must conform to the code requirements for its occupancy, not for the more hazardous occupancy. Finally, statement IV is incorrect. Occupancy separations are of one-, two-, three-, or four-hour fire-resistive construction, as required by the code. Only statement II is correct, making D the right answer.

4. An interior partition is constructed of metal studs 24" on center with two layers of 1/2" Type "X" gypsum wallboard on each side. What is the fire-resistive rating of the partition?

   A. None
   B. 1/2 hour
   C. 1 hour
   D. 2 hours

Fire-resistive ratings for various walls and partitions are shown in Table 43-B of the Uniform Building Code. According to this table, the partition described has a two-hour rating (correct choice D). Is this a fair and realistic question? Are you actually expected to memorize fire ratings of various wall assemblies? No, you cannot be expected to commit all that information to memory. But you should know where to find fire-resistive ratings, should the code or portions of it be available during the exam. And you should also have a general idea of the range of fire ratings possible using various materials.

5. If the maximum allowable floor area of a one-story building having a certain occupancy and type of construction is 12,000 square feet, the total allowable area of all floors of a four-story building of the same occupancy and construction type is

   A. 12,000 sq. ft.
   B. 24,000 sq. ft.
   C. 36,000 sq. ft.
   D. 48,000 sq. ft.

The Uniform Building Code specifies that the total area of all floors of multistory buildings shall not exceed twice the area allowed for one-story buildings. Thus, in this question, the total area of all floors is $2 \times 12,000 = 24,000$ square feet (answer B). Note that the total allowable area of all floors is the same, whether the building has 2, 3, 4, or more floors.

6. You are the architect for a two-story office building 100 feet by 100 feet in size. Select the correct statements concerning exit requirements for the building.

   I. At least two exits are required.
   II. Each exit must be at least 32 inches wide.
   III. The exits must be at least 71 feet apart.
   IV. The maximum distance from any point in the building to an exit is 75 feet.

   A. All of the above
   B. I, II, IV
   C. III
   D. I, II, III

Exit requirements are governed by Chapter 33 of the Uniform Building Code. From Table 33-A, the total occupant load for this office building is the area in square feet divided by 100 square feet per occupant. So, the occupant load is $2(100 \times 100)/100 = 200$. Since this is greater than 30, a minimum of two exits are required (I). From Section 3303, the clear width of an exitway must be at least 32 inches (II). The arrangement of the exits is determined by Section 3302. The minimum distance apart is one-half the diagonal dimension of the building $= (1.41 \times 100)/2 = 71$ feet (III). Finally, the maximum distance from any point to an exit is 150 feet, not 75 feet as given in IV. Since I, II, and III are correct, the answer is D.

7. Which of the following statements would least likely be found in the building code?
   A. Stairs in smokeproof enclosures shall be of noncombustible construction.
   B. No leaf of an exit door shall exceed four feet in width.
   C. The maximum height and number of stories of a building shall be dependent on its occupancy and type of construction.
   D. The area of the building may not exceed 75% of the total area of the building lot.

Candidates often find it difficult to differentiate between a building code and a zoning ordinance. The primary purpose of the building code is to provide for the safety and health of the public. It is created by an ordinance to regulate the construction of buildings within a municipality. In contrast, zoning is created by an ordinance to regulate the character and use of a parcel of land. Zoning ordinances stipulate the type of development; i.e., residential, commercial, industrial, agricultural, and so forth, that may take place within a given area. In most instances, zoning ordinances also control the extent of allowable land coverage, setbacks from property lines, allowable densities, and building heights and areas. Building codes also regulate maximum building heights, areas, and allowable number of stories. However, these limitations are based on construction and occupancy types to provide for the safety of the public. Similar restrictions imposed under zoning ordinances are intended to provide uniform standards of development to protect the environment and property values. The correct answer is D.

8. Glazed openings in one-hour fire-resistive corridor walls would most likely be
   I. 1/4" thick tempered glass.
   II. 1/4" thick wired glass.
   III. installed with wire clips.
   IV. installed in steel frames.
   V. prohibited by most building codes.
   A. II and III
   B. I and IV
   C. II and IV
   D. V only

The Uniform Building Code stipulates certain minimum requirements for door, window, and duct openings in one-hour walls separating interior corridors from adjacent spaces, in order to ensure safe egress for occupants in case of fire. Doors must be protected by smoke and draft control assemblies and must have a label showing the rating. Furthermore, the doors must be automatic-closing or self-closing. All other interior openings must be fixed and protected by 1/4" wired glass in steel frames (correct answer C). Tempered glass is commonly required in glazed areas subject to impact, such as sliding glass doors. Wire clip installations are allowed for glazed openings in casement windows where these are employed for 3/4 hour fire windows.

9. A building's occupant load is determined by
   A. the floor area and use of the building.
   B. its location in a Fire District created by ordinance.
   C. its location on the property.
   D. the fire-resistive rating of its exterior walls.

A building's occupant load is the total number of persons that may occupy the building or any portion thereof at any one time. The occupant load that is permitted in any building is calculated by dividing the floor area assigned to a specific use by the square feet per occupant as designated by the building code. The available square feet per occupant is usually determined by the building's use. Consequently, if the available area per occupant for offices is 100 SF, a 10,000 SF office building would yield an occupant load of (10,000 ÷ 100) = 100 persons. A building's location in a Fire District establishes restrictions concerned with the fire resistance of its construction. A building's location on the property—that is, the distance from its exterior walls to adjacent property lines and streets—establishes fire ratings of exterior walls, numbers, sizes and location of exterior openings, and so forth. The correct answer is A.

CITY & SITE PLANNING

1. The soil report for a project discloses that the soil beneath the site is stable at its natural moisture content, but swells upon wetting. In view of this condition, which of the following are the most correct statements?

   I. A structural first floor should be provided, supported by a foundation system extending through the zone subject to wetting.
   II. The foundation system should consist of conventional spread footings, using a very low soil bearing value.
   III. Basement walls should be provided with weep holes to relieve hydrostatic pressure.
   IV. Unless positive drainage is provided, basement walls should be designed for an equivalent fluid pressure greater than 30 pounds per cubic foot.

   A. I, IV
   B. I, II, III
   C. II, IV
   D. All of the above

   Here we are required to select the correct statements based on the sketchy information furnished. In areas where swelling soils are present, a structural floor with an air space beneath, supported by a foundation system not subject to movement, is the best means of minimizing the risk of floor slab heave. I is therefore correct. Statement II is incorrect for a couple of reasons. First of all, we are not given sufficient information to determine if, indeed, the foundation system should be conventional spread footings. Furthermore, using a low soil bearing value in swelling soils is not desirable, since the upward pressure of the expanding soil can easily exceed the low downward pressure of the structure and hence cause foundation heaving. Part of III is correct: hydrostatic pressure may be exerted on basement walls and some means of relieving this pressure should be used. But weep holes are not a good solution, since they would allow water and dirt to flow into the basement. III is therefore false. Finally, statement IV is true. Unless a subsurface drainage system around and under the basement is provided, swelling soils can exert pressures against surfaces they contact far in excess of an equivalent fluid pressure of 30 pounds per cubic foot. Since only I and IV are correct statements, A is the right answer.

2. Parcels of land are described by a variety of methods. In this regard, select the *incorrect* statement from those which follow.

   A. A metes and bounds description consists of a starting point, followed by the successive lengths and bearings of each boundary line.
   B. The national system of surveying public lands is based on townships six miles square.
   C. Urban land is often described by lot, block, and tract names or numbers.
   D. Coordinate systems are not generally used, because such systems neglect the earth's curvature, resulting in substantial errors.

   Legal descriptions of land may take any of several forms. The metes and bounds method consists of a narrative description of a point of beginning, followed by a statement of the length (mete) and bearing (azimuth) of each boundary line, continuing around the parcel until the point of beginning is reached. A is therefore a correct statement. The U. S. system of surveying public lands, which started in 1785, is based on townships six miles square, which are divided into 36 square sections, each of which is one square mile or 640 acres in size. Sections may be further subdivided into quarters and quarter-quarters (sixteenths). B is also correct. C is also a correct statement. A parcel of land within a city is often identified as a specific lot of a particular block in a tract, for which an official record map exists. In certain cities or metropolitan districts, all points are referred to a system of plane rectangular coordinates, using some point as the origin of the coordinate system. While such a system disregards the curvature of the earth, the resulting error is found to be extremely small. D is the only incorrect statement and therefore the answer to this question.

3. All of the following information may be found on a topographic map, *except*
   A. property lines, easements, and utilities.
   B. location of streams.
   C. location of roads and buildings.
   D. location and identification of soil conditions.

The usual topographic map shows the location, size, and shape of physical and other features of a piece of land, including property lines, easements, and utilities (A); natural features, such as streams (B); man-made features, such as roads and buildings (C); and contours indicating elevation. Soil conditions, however, are not generally shown on a topographic map (correct answer D).

4. Under normal circumstances, a slope of 10 percent is a desirable maximum for which of the following uses?
   I. Lawns
   II. Paved walkways
   III. Drainage ditches
   IV. Planted banks (unmowed)
   V. Unretained earth cuts

   A. II, III   C. I, IV, V
   B. I, II, IV   D. II, III, IV

First of all, let's define percentage of slope. A one foot vertical rise in a horizontal distance of ten feet is a 10 percent slope (1/10 = 0.10 = 10%). With that in mind, the following grades are generally accepted standards. Lawns (I) should be kept under 25 percent to facilitate mowing and maintenance. The maximum slope for paved walkways (II) is usually 10 percent, but may be 12 percent for very short ramps. Drainage ditches (III) vary in slope from a minimum of 2 percent to a maximum of 10 percent. Finally, planted banks and unretained earth cuts (IV and V) are usually limited to a 50 percent grade, since slopes steeper than this are highly susceptible to natural soil erosion. The correct combination of answers (II and III) is found in answer A.

5. Humus is the
   A. well-decomposed, more or less stable part of the organic matter in mineral soils.
   B. natural layer of plant residues or other materials on the surface of the soil.
   C. highly decomposed organic soil material developed from peat.
   D. mass of rotted organic matter made from waste plant residues.

Questions pertaining to landscaping may test the candidate's understanding of soils, soil improvement, fertilizers, and various organic and inorganic matter utilized in growing plant material. The definitions listed above refer to various matter concerned with this process. Choice A, the correct answer, defines humus. Choice B refers to "mulch" which is generally used to help conserve moisture, control temperature, prevent surface compaction, reduce runoff and erosion, improve soil structure, or control weeds. Choice C refers to "muck" that is a combination of soil and water, having a higher mineral content than peat. It is decomposed to the point where the original plant parts cannot be identified. Choice D refers to "compost" that, when mixed with nitrogen and soil, is used as an organic fertilizer. The principal purpose in making compost is to permit the organic materials to become crumbly and to reduce the carbon-nitrogen ratio of the material.

6. Foundation material that tends to swell and shrink with changes of moisture content is referred to as expansive soil. Which among the following types of soil is considered expansive?

   I. Sand
   II. Gravel
   III. Silt
   IV. Clay
   V. Rock

   A. I and III
   B. II and V
   C. III and IV
   D. IV

Candidates should be acquainted with the general properties of soils listed in this question, as each has differing implications for site planning. Silts and clays (III and IV) swell when they become wet and are referred to as expansive soils (correct answer C). Such soils will exert upward pressures on footings and slabs when the moisture content increases and can cause considerable damage to buildings. Sand and gravel (I and II) are well drained and make good foundation material. Rock (V) has the highest bearing capacity of any material, and footings found on rock are usually smaller than those founded on soil.

7. Which of the following statements are correctly matched with the vehicular circulation systems they describe?

   I. Takes advantage of the natural topography.
   II. Generally best on flat or slightly rolling land.
   III. Often results in visual monotony.
   IV. Lacks flexibility because its center is fixed.
   V. Connects flow between two points.
   VI. Streets are more interesting because of variations in views.

   A. Curvilinear System—I and VI
   B. Grid System—II, III, V
   C. Radial System—IV and V
   D. Linear System—II only

Circulation systems are arbitrary patterns that have developed through history to solve specific traffic problems. The curvilinear system (A), for example, utilizes the natural topography (I) by following the contour of the land as much as possible. In this system, there are fewer through streets than in the grid system. Cul-de-sacs and dead end streets are commonly associated with the curvilinear system, and streets are more interesting (VI) because of the different views, street types and lengths, and the *adaptability to topographic variations. These accurately matched statements make A the correct answer. The grid system (B) usually consists of equally spaced streets which run perpendicular to each other. This system of circulation is most often used on flat or gently sloping land (II). Because of its rigid geometric pattern, it often results in visual monotony (III), lacking regard for the natural topography. The radial system (C) directs flow to a common central point. In cases where there is a great deal of activity at the center, however, it may cause congestion and crowding. Since the center is a permanent point, it becomes difficult to make changes in the pattern (IV), in contrast to the grid system, for example. The linear system of circulation (D), such as that of railway or canal systems, connects traffic flow between two or more points (V). If the movement along its length is overloaded, traffic becomes congested and the flow is impeded.*

8. Which of the following trees would be best suited to act as a screen against prevailing winter winds?
   I. Engelmann Spruce
   II. Cottonwood
   III. Quaking Aspen
   IV. Scotch Pine
   V. American Beech
   A. I and IV
   B. II and III
   C. III and IV
   D. III and V

Site vegetation is able to influence the thermal environment of a building by diverting storm winds, channeling cooling summer breezes, and shading the building from the sun. In order to be effective in north temperate climate conditions, wind screens require a dense massing of trees. Evergreens are ideally suited for this purpose. In contrast, deciduous trees have little screening value in the winter, but are used for sun shading in summer because of their dense foliage. This question about environmental control through the use of landscaping tests whether we understand the characteristics of some common types of evergreen and deciduous trees. Since I and IV are the only evergreens appearing in the list, the correct answer is A.

9. Match each of the following terms with its correct meaning.
   I. Percolation
   II. Erosion
   III. Creep
   IV. Infiltration
   1. The slow mass movement of soil down steep slopes under the influence of gravity, aided by water and alternate freezing and thawing.
   2. The wearing away of the land surface by running water, wind, or ice.
   3. The movement of gravitational water through soil.
   4. The movement of groundwater or hydrothermal water into rock or soil through joints or pores.
   A. I-3, II-2, III-4, IV-1
   B. I-3, II-2, III-1, IV-4
   C. I-2, II-3, III-1, IV-4
   D. I-1, II-4, III-3, IV-2

The effects of wind and water on land are considerable. Each of the terms above is related to these effects, and the terms are correctly matched in answer B.

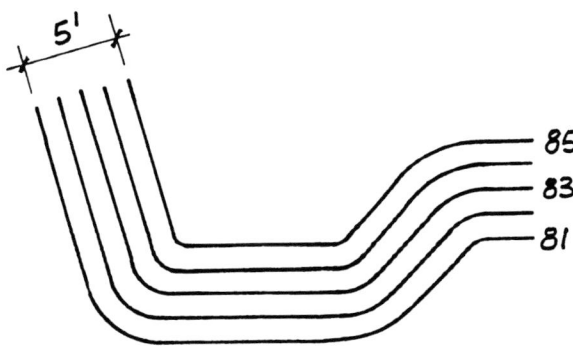

10. For the graded bank shown above, the grade of the slope is

A. 12.5 percent   C. 80.0 percent
B. 40.0 percent   D. 125 percent

Many candidates find it confusing to analyze the slope of the ground. However, it is important to be able to utilize the topography of a site in order to maximize the use of available land, both in site planning and building design. Grades for existing as well as man-made slopes as shown above are established by measuring the distance between contours at a given scale and a given contour interval. The formula is

% grade =

$$\frac{contour\ interval}{distance\ between\ contours} \times 100$$

In the graded bank above we are shown a four foot change in elevation for every five horizontal feet of land. Using the formula we calculate as follows:

$$\%\ grade = \frac{4}{5} \times 100 = 80\%$$

The correct answer is C.

CONCRETE

1. Select the *incorrect* statement.

   A. The principal reason for using entrained air in concrete is to improve its resistance to freezing and thawing.

   B. The strength of concrete increases as the water-cement ratio decreases.

   C. If formwork must be removed as quickly as possible, it is advisable to use Type III portland cement in the concrete mix.

   D. During cold weather, calcium chloride should be added to lower the freezing point of concrete.

   In this question, we are looking for the one incorrect statement. A and B are both true. Choice C is also correct; since Type III cement provides high strength at an early age, it is often used when forms must be removed quickly. D is the incorrect statement and therefore the answer to this question. High strength at an early age is often desired during cold weather, to reduce the length of time weather protection is required. Since calcium chloride is the most common accelerating admixture, it is frequently used for this purpose. However, using calcium chloride to lower the freezing point of concrete should never be permitted, because the quantity needed is so great that strength and other properties would be seriously affected.

2. The following illustration is a typical example of

   A. flat slab construction.

   B. flat plate construction.

   C. lift slab construction.

   D. precast concrete plank system.

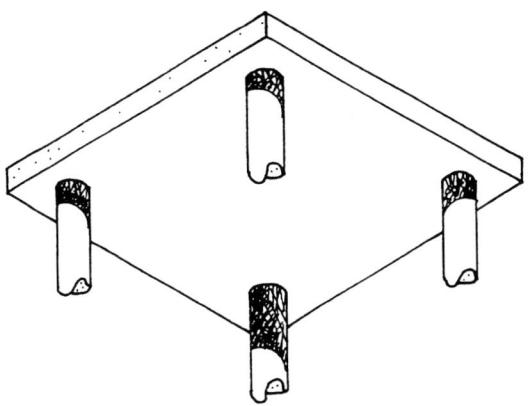

   The term "flat slab" (choice A) refers to a reinforced concrete floor system with no beams or girders, but having drop panels and column capitals, neither of which are shown in the sketch. Drop panels are thickened slabs at columns and column capitals are flared haunches at the tops of columns. Girderless floors without drop panels or column capitals are called flat plate floors (choice B). The simple formwork required for this system offers obvious economy. The lift slab system (choice C) is a precast flat plate system in which the floor and roof slabs are cast at ground level, one over another. The slabs are raised to their final elevations by means of hydraulic jacks. Precast concrete planks (choice D) are precast solid one-way slabs which span between beams. Thus, the sketch might conceivably represent either the flat plate (B) or lift slab (C). However, since flat plate is the more inclusive term, and since nothing in the sketch specifically identifies it as lift slab, B is the preferred answer.

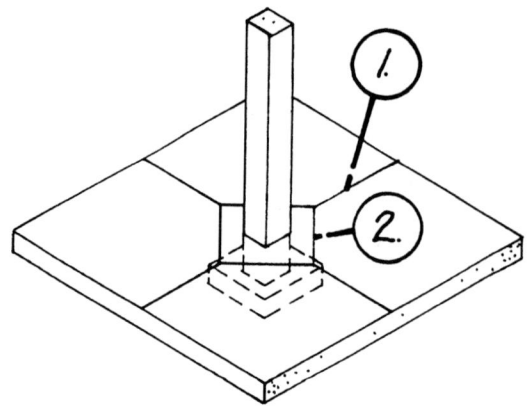

3. Shown above is a section of a concrete slab on grade with an interior column. The pattern of joints shown is used

   A. mainly for aesthetic reasons.
   B. because it is impossible to tool a joint neatly directly adjacent to the column.
   C. so that the footing will be free to settle without causing major cracking of the slab.
   D. because surface cracks generally occur on a 45 degree angle.

4. With reference to the same detail above, the joints designated 1 and 2 are known respectively as

   A. control joint and keyed joint.
   B. control joint and isolation joint.
   C. expansion joint and control joint.
   D. expansion joint and construction joint.

Joint 1 is a control joint, used to eliminate random cracking caused by drying shrinkage of the concrete slab. Joint 2 is an isolation joint which separates the slab from the column and permits the column footing to settle independently of the slab. The correct answers are 3-C and 4-B.

5. Excessive water rising to the surface of a freshly cast concrete slab is commonly known as

   A. efflorescence
   B. laitance
   C. bleeding
   D. honeycombing

Efflorescence (A) is the whitish powder of crystallization which comes to the surface when water evaporates from brick. Laitance (B) is a low-strength layer of fine particles that floats to the surface of wet concrete. Honeycombing (D) refers to the voids and cavities on a concrete surface caused by insufficient fine aggregate or inadequate concrete compaction. The correct term is bleeding (C).

6. A slump test is used to determine the

   A. compressive strength of concrete.
   B. tensile strength of concrete.
   C. workability of concrete.
   D. quality of the aggregate in the concrete mix.

A slump test is used to determine the consistency and workability of concrete (correct answer C). Stiff mixes have low slump, while more fluid mixes have greater slump.

7. During your observation of the reinforcing steel placement on a construction project, you see a reinforcing bar with the following markings on its surface: H 7 N 60. What does this tell you about the bar?

   A. It is a 7/8 inch diameter bar made of new billet steel and conforming to grade 60.
   B. It is a 7/8 inch diameter bar made of new billet steel and conforming to grade H.

C. It is a 7/8 inch diameter heavy-duty bar conforming to grade 60.
D. It is a 3/4 inch diameter bar made of new billet steel and conforming to grade 70.

You are expected to know how to identify various construction materials, and this kind of question has appeared on past exams. During the rolling process, identification markings are embossed on the surface of reinforcing bars. In this example, "H" identifies the producing mill, "7" denotes the bar size (7/8 inch), "N" refers to the type of steel (new billet), and "60" refers to the grade of steel (grade 60 has a minimum specified yield stress of 60,000 psi). The correct answer is therefore A.

CONSTRUCTION SITUATIONS

1. You are the architect for an auditorium facility, which contains an underground concrete utility trench. The contractor asks you to provide him with shoring details for the earth excavation required to construct the trench. What should be your response?

   A. Comply with the contractor's request.
   B. Advise the contractor that your mechanical engineering consultant is responsible for the shoring.
   C. Refer the contractor to the local Public Works department, which is responsible for all underground construction.
   D. Inform the contractor that he is solely responsible for construction methods.

   The General Conditions (AIA Document A201) clearly state that the contractor, and not the architect, is solely responsible for all construction means, methods, techniques, sequences, and procedures. D is therefore the correct answer.

2. An existing covered walk is accidentally damaged by a piece of grading equipment during construction of a school project. Who is responsible for the cost of replacement of the covered walk?

   A. The general contractor
   B. The equipment operator
   C. The owner
   D. The grading subcontractor

   The General Conditions (AIA Document A201) state that "the contractor shall be responsible to the owner for the acts and omissions of his employees, subcontractors and their agents and employees, and other persons performing any of the work..." The correct answer is therefore A.

3. The working drawings for a project show lateral bracing of the suspended ceiling for earthquake resistance. Although the architect did not inspect the ceiling construction before it was covered up, he suspects that the lateral bracing was not properly installed. The building inspector has not noted any violations. What should the architect do?

   A. Nothing, since no violations have been noted by the inspector, and the architect has no evidence that the construction is not in conformance with the contract documents.
   B. Prepare a change order requiring the contractor to uncover the ceiling for the architect's inspection.
   C. Request the contractor to uncover the ceiling for the architect's inspection.
   D. Ask the owner to direct the contractor to uncover the ceiling, since the architect lacks such authority himself.

   Under the circumstances described, the architect may request to see the concealed construction, and it must be uncovered by the contractor (correct choice C). This situation is covered by the General Conditions (AIA Document A201) Paragraph 13.1.2. It should be noted that the responsibilities of the building inspector are different from those of the architect, and the architect should never rely on the inspector as a substitute for his own on-site observations. Incidentally, if the uncovered work is found to be in accordance with the contract documents, the cost of uncovering and replacement is borne by the owner. On the other hand, if the work does not conform to the contract documents, the contractor must pay the costs.

4. During the excavations for footings for a large office building, the contractor determines that the engineering soils survey and report neglected to indicate a substantial area of loose fill soil. After inspecting the area, the structural engineer orders the contractor to remove the fill and replace it to a compacted density of 90 percent where the building footings and slabs are affected. The contractor requests a change order reflecting an addition to the contract sum, as well as an increase in the contract time required by the additional excavation and compaction. The architect claims that it is the contractor's responsibility to acquaint himself with the conditions of the site whether or not the soils report or contract documents indicated unusual conditions below grade. In accordance with standard accepted procedure, the

   A. contract sum and time should be adjusted by Change Order.
   B. architect is correct in his assumption and the contractor must bear the additional cost.
   C. soils engineer is liable for the additional cost and should negotiate a settlement with the contractor.
   D. architect should issue a Written Order to effect the change.

The General Conditions of the Contract for Construction, AIA Document A201, provides the following: "Should concealed conditions encountered in the performance of the Work below the surface of the ground, or should concealed or unknown conditions in an existing structure be at variance with the conditions indicated by the Contract Documents; or should unknown physical conditions below the surface of the ground, or should concealed or unknown conditions in an existing structure of an unusual nature, differing materially from those ordinarily encountered and generally recognized as inherent in work of the character provided for in this Contract, be encountered, the Contract Sum shall be equitably adjusted by Change Order upon claim by either party made within 20 days after the first observance of the conditions." The correct answer is A. Written Orders are issued by the architect and indicate minor changes in the work without change in contract sum or contract time.

5. Six months after completion of a project, the owner notifies the architect that the cooling cycle of the air conditioning system is unable to cope with the heat gain in the building, resulting in temperatures above the comfort zone. Upon careful investigation, the architect's mechanical engineering consultant determines that the mechanical subcontractor did not install the air handling units in accordance with the drawings and specifications. Which is the most likely course of action the architect would follow?

   A. Advise the owner to submit a written notice to the general contractor asking him to make the necessary correction to the work.
   B. Try to work with his consultant to find a means to modify the system in order to provide the necessary cooling.
   C. Sue the mechanical engineering consultant for lack of supervision on the job.
   D. Recommend that the owner seek the funds from the general contractor to make the necessary changes to the system.

When it becomes apparent after completion of construction that the general contractor or any of his suppliers or subcontractors failed to comply with the requirements of the contract documents, the architect should advise the owner to request the necessary corrective work from the general contractor (A). It is not incumbent upon the architect to find ways to correct the general contractor's work, even if the defective work does not come to his attention until after completion (B). Nor can he sue his consultants, since supervision is not the responsibility of either architect or consultant (C). The owner should, under no circumstances, assume the responsibility to correct any work with his own forces, even if the general contractor agrees to furnish the necessary funds for this purpose (D). The responsibility for the correct functioning of the cooling system must remain that of the general contractor until all contractual responsibilities are met. Only if and when the general contractor refuses to comply should a suit to recover damages be considered. And since the work is still within the one year guarantee period, the owner is in a very strong position to demand the necessary corrective action, although the statute of limitations would, in fact, extend this right beyond that period. The correct answer is A.

6. A hotel building is substantially completed except for the installation of finish hardware in the guest rooms. In the interim, the contractor has installed temporary hardware and has submitted a letter to the owner stating that the finish hardware will be installed as soon as it is delivered to the job site by the distributor. The contractor asks the owner to accept the building in order to receive final payment, thereby reinstating his bonding capacity to bid other projects. The architect should advise the owner to do which of the following?

A. Request a check from the contractor in the full amount of the hardware and its installation prior to making final payment.

B. Withhold final payment until all the finish hardware is in place and functioning properly.

C. Demand a bond covering the installation of the hardware and subsequently make the final payment to the contractor.

D. Terminate the agreement, subtract the cost of the hardware and its installation, make the final payment, and subcontract the work to another contractor.

The process leading to final completion and final payment begins when the contractor requests the architect to issue a certificate of substantial completion. Uncompleted items of work are recorded on a "punch list," and a certificate of substantial completion is issued. When all items on the punch list are completed, the contractor requests final inspection and submits his final application for payment. If the work is found acceptable and complete, the architect issues the final certificate for payment which includes the release of the contractor's retainage, that is, the sum held back from each progress payment. Provision is normally included in the General Conditions to cover cases where final completion of the work is materially delayed through no fault of the contractor. This permits payment to be made for that portion of the work that is completed and accepted, but it does not terminate the contract. In the situation described above, a substantial amount of work remains uncompleted. Since final payment by the owner constitutes a waiver of all claims by him except those arising from unsettled liens, etc., the architect would expose the owner and himself to potential liability if he were to advise the owner to do anything but withhold final payment until all the work included in the contract is completed and accepted. The correct answer is B.

The following situation applies to questions 7, 8, and 9.
Specifications for the carpet to be installed in a school require a specific quality and design as manufactured by Pyle Carpets. During the architect's color presentation for the owner's approval, the owner instructs the architect to substitute a cheaper carpet, manufactured by Loomo Carpets, because of his preference for a particular color not available in the Pyle line. The architect requests a credit from the contractor in order to issue a Change Order. The Loomo carpet is limited in supply because it is no longer in production, and therefore, the architect instructs the contractor to place a hold for the exact amount of carpet to insure its availability at the time of installation. The requested credit is not furnished, nor is the Change Order issued. Three months later, the contractor informs the architect that he cannot furnish the Loomo carpet because it is no longer available. The owner refuses to accept a carpet from the Loomo Company because none of the other colors are acceptable to him. To expedite the work, the owner approves a carpet from Weevo Carpet Company in the same price range as the originally specified carpet. The contractor agrees to order the material and requests an increase in the contract price.

7. The architect disallows the increase because
   A. the contractor is at fault for not ordering the substituted material when instructed.
   B. the owner refuses to sign a Change Order to increase the contract price.
   C. the carpet finally selected does not cause an increase in cost to the contractor.
   D. the contractor did not issue a Change Order reflecting the substitution.

In construction situations where substitutions are requested by the owner, the architect should obtain a quotation for either an increase or a decrease in the contract amount and issue a Change Order calling for the substitution and the amount of the change in cost. In this case, the contractor neither furnished a credit nor did the architect issue a Change Order; consequently, the contract amount remained the same as the original bid amount. Since the carpet ultimately substituted for the original did not cause an increase in the cost to the contractor, the architect would not allow an increase. The correct answer is C.

8. The responsibility to assure the eventual availability of the carpet no longer in production is that of the
   A. Architect
   B. General Contractor
   C. Owner
   D. Flooring Contractor

Once instructions are given for changes on the job it becomes the general contractor's responsibility to effect these changes. If, under certain conditions, he is unable to comply with the owner's or architect's request, he must immediately inform the architect and await further instructions. In no way does the architect or the owner assume responsibilities in the procurement of materials specified in the contract documents. Neither, for that matter, are the subcontractors, in this case the flooring contractor, in any way directly responsible to the owner. The correct answer is B.

9. In order to avoid a similar situation in the future
   A. the architect should specify a carpet in production.
   B. the owner should furnish the carpet for installation by the contractor.
   C. the architect should advise the owner to select a substitute that is available.
   D. the contractor should be instructed to store the substituted carpet on site until installation.

It is virtually impossible during the preparation of specifications to foresee the eventuality of the discontinued production of an item. It is not uncommon for materials to be out of stock or production at the time they are required on the job. Substitutions are common and must be dealt with efficiently and equitably. To assure the availability of a material when a change is made during construction, the architect should recommend a material that is readily available, and one that will not hold up the job because of discontinuance or unavailability. The correct answer is C.

DESIGN THEORY

1. Which of the following statements, concerning design standards, has the *least* validity?

    A. Areas with higher than normal ceilings are best located on the ground floor.

    B. Areas that require sloped floors are best placed directly on grade.

    C. Spaces containing plumbing fixtures are best arranged adjacent to one another.

    D. Spaces programmed for children are best situated on the level that connects most directly to the street.

All of the general statements are more or less true; however, it is correct answer A that is only partially right and, therefore, least valid. Areas that have high ceilings are often placed on the ground floor because the higher ceiling implies a large space that is often used for many people. It is also true, however, that high-ceiling areas are appropriately placed on the highest floor where the roof level can be modified without affecting the rest of the building. This is often the case where mechanical equipment is involved, and also suitable where different ceiling levels or sloping ceilings are required. Statements B and C are nearly always true, as good design sense would probably indicate. With regard to choice D, it is common to locate spaces for children on the ground floor in order to avoid steps, which are often considered a safety hazard. Ground level access also provides an efficient means to pick up or deliver children, whether by bus or private car.

2. Which of the following statements is *false*?

    A. Anything that can be seen has shape.

    B. All shapes have size.

    C. All shapes have mass.

    D. All shapes have texture.

In order to answer this question candidates must be able to distinguish among the terms describing various elements of design. It is true, for example, that shape is the outline or configuration of anything that can be seen (A). Proceeding from that point, all shapes possess physical dimensions, or size (B); and they also possess surface characteristics or texture (D), whether they are glassy smooth, very rough, or anything in between. However, not all shapes have mass (correct answer C); this quality applies only to three-dimensional volumes that have density or bulk. Two-dimensional shapes have no mass.

3. Which of the following schematic diagrams of a hypothetical office structure shows the most direct vertical access from the ground floor level of the building to the tower?

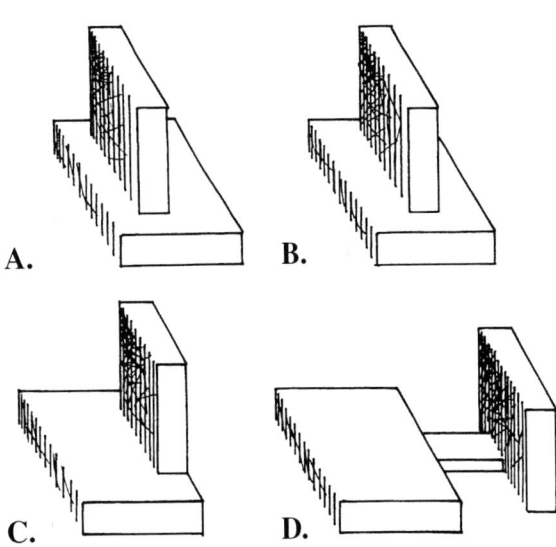

Questions such as this are difficult to answer because it is often unclear what is being asked. We must first determine what "direct vertical access" really means; and since there is very little to go on, we must proceed with the analysis by making one assumption

43

at a time. We assume that "vertical access" refers to elevators located at the center of the tower. Next we assume that "direct" refers to the circulation path to these elevators from any part of the ground floor. Therefore, we are looking for the diagram that places the tower centrally over the ground floor, so that access from any point is relatively equidistant. The correct solution is answer B, since the other diagrams show greater distances from the farthest ground floor points to the center of the tower.

4. Which of the following functional spaces generally demands the most distinctive architectural form?

   A. A banquet room
   B. An auditorium
   C. A museum gallery
   D. An industrial exhibit space

Various building types have been designed in a wide variety of shapes, based not only on function, but also on the designer's arbitrary preferences. Among the functional spaces listed, all can be accommodated in an essentially rectangular shape except for an auditorium (correct answer B). Auditoriums often require sloped or curved walls and ceilings for acoustical reasons, and they invariably have sloped floors to provide unobstructed sight lines. The other functions listed require a flexible space in which a diversity of layouts can easily be accommodated.

5. Diagrammatic layouts, also referred to as concept or bubble diagrams, graphically represent which of the following?

   I.   Floor area comparisons
   II. Circulation relationships
   III. Structural considerations
   IV. Physical configurations
   V.   Functional associations

   A. I and II   C. I, III, V
   B. II and V   D. I, IV, V

The principal purpose of a diagrammatic layout is to determine graphically the functional relationships (V) among various programmed spaces. A quick concept diagram shows each space as a rough shape with no regard to scale. These shapes are then connected by lines or arrows to other shapes with which they have a functional association. By so doing, one establishes not only a graphic indication of the functions, but also the important circulation relationships (II) that exist among the spaces. Thus, diagrammatic layouts deal with circulation and function, as stated in correct answer B. Bubble diagrams do not deal with areas (I), since they are not to scale; nor do they involve structure or form (III and IV), which are generally considered later in the design process.

6. In a structure located in the southwest desert area of the U.S., which of the following design features would most significantly recognize the climatic problems of the area?

   A. Insulated glass windows
   B. Flat roof planes
   C. Fixed vertical louvers
   D. Deeply recessed openings

The southwest desert location implies that the overwhelming climatic problem in the structure is solar heat gain. In this regard, we can eliminate the flat roof feature (B), since roof shape alone has little effect in reducing heat gain. Insulated glass (A) windows are helpful, and this is a possible answer. Fixed vertical louvers (C), especially on the south side of a building, do little to diminish direct solar heat gain when the sun is high. Horizontal louvers would be more efficient, but that is not one of our choices. Finally, deeply recessed openings (D) are very effective in shading glass, regardless of their orientation. In a choice between insulated glass and shaded glass, shaded glass (correct answer D), is preferred because preventing sun from reaching the glass is always more effective in reducing solar heat gain than controlling the gain once the glass has been exposed to sun.

7. The most efficient functional design in the majority of buildings results when corridors

   A. are eliminated.
   B. divide areas of similar function.
   C. occur on each side of functional areas.
   D. border functional areas.

The meaning of efficent design may be controversial, but in most cases it results when corridors are arranged alongside functional areas that they serve (correct answer D). Can we assume then that the design is twice as efficient when two such circulation paths are used (C)? Probably not, since most spaces do not require double the amount of circulation space, and if two corridors were used, the placement of exterior openings might be restricted. The elimination of all corridors (A) would certainly save floor area, but circulation through functional areas would be very disruptive. Finally, dividing functional areas with corridors (B) would be even less efficient than eliminating all corridors, since it would seriously impair the function of the divided space.

8. The following schematic designs represent a typical floor of a multi-story apartment building located along a body of water. Which design would best accomplish the objective of providing a view of the water? Assume an equal number of units in each design.

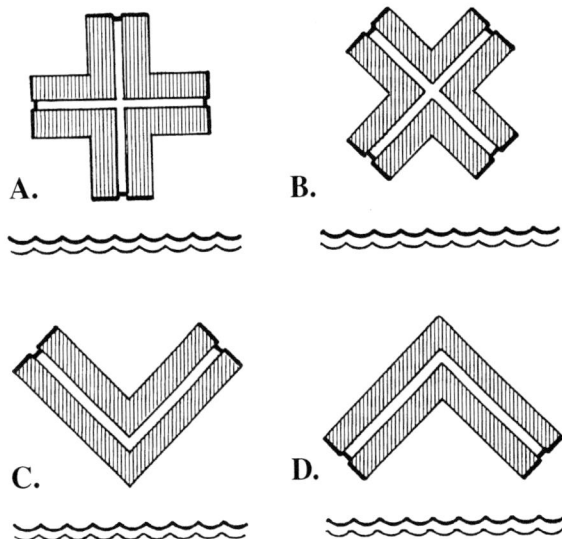

In this question we must assume that complete satisfaction of the objective is impossible, since every side of the structure cannot be oriented toward the water. Therefore, the best solution will be the one that allows the greatest number of units to overlook the water. We should also assume that an authentic view is one that encompasses up to a 45 degree angle from any unit. In design A, exactly half of the units have a view of the water. By rotating the cruciform plan, as in design B, we have not materially changed the relationship of the units to the

water; half of the units still have a view. In schematic designs C and D we have an L-shaped structure in which more units can be placed on one side of the corridor than the other. Therefore, the solution in which the long sides face the water (correct answer C) is the one in which the greatest possible number of units have a water view. In fact, design C provides a water view for more than half of the units, while design D provides a view for somewhat less than half of the units.

9. Select the combination of correct statements from among those that follow.
   I. Color can affect one's perception of time.
   II. Color can cause psychological stress.
   III. Color and light involve two different perceptions.
   IV. Color exists in all things at all times.

   A. I and II C. I, II, III
   B. III and IV D. I, II, III, IV

Color is the appearance of something caused by the quality of light that is reflected by it. Therefore, color and light are inseparable; one cannot exist without the other, and hence statement III is false. Since color exists only when there is enough light to perceive it, it follows that color ceases to exist in the dark, and thus, statement IV is also false. By elimination we are left with the correct answer A, which includes two statements regarding the inherent power of color. First of all, color can affect one's perception of time; for example, in a green environment time seems to pass more slowly than in a red one. Secondly, color can affect one's psychology, since all people respond emotionally to color. Red, for example, is considered fiery and exciting, green is soothing, and black is mysterious. Used in a reckless way, color can cause not only psychological stress, but it can actually make one physically ill.

10. In the past several years there have been a number of design solutions for tall buildings which employ twin towers. Which of the following reasons is the most reasonable justification for such a design scheme?
    A. Twin towers provide greater aesthetic distinction and visual impact.
    B. Twin towers provide greater opportunity for original design expression.
    C. One very high tower may require an unreasonable amount of floor area to be devoted to vertical circulation.
    D. One very high tower may be unreasonable in size or structurally unwieldy.

From the World Trade Center in New York to Pennzoil Place in Houston, twin tower high-rise schemes have proliferated throughout the country during the past decades. There may be several reasons for these developments, but from a standpoint of design theory, twin towers make sense when a single tower simply becomes too large (correct answer D). That was Yamasaki's rationale for the World Trade Center, in which he had to accommodate nine million square feet of office space. Twin towers can be visually distinctive (A) and provide an opportunity for originality (B), but there is no reason why a single tower, for example, Frank Lloyd Wright's proposed Mile High Skyscraper, cannot be equally as dramatic or unique. Finally, the problem of floor area consumed by vertical circulation has been solved recently by using stacked elevator shafts with sky lobbies and even double-decked elevator cabs, so that this kind of space has been kept within reasonable limits.

DOCUMENTS

1. A certificate for payment is
   A. a draft against the owner's bank account, payable to the contractor for work completed.
   B. a statement by the architect informing the owner of the amount due the contractor.
   C. a form submitted by the contractor to the owner, requesting payment for work completed, less retainage.
   D. a form submitted by the architect to the owner, requesting payment for architectural services rendered.

   Upon completion of a portion of the work, the contractor submits an application for payment to the architect, who determines the value of work performed and transmits a certificate for payment to the owner for payment (correct answer B). For this purpose, the AIA provides a combined Application and Certificate for Payment (AIA Document G702), which shows the total dollar amount of work completed and materials stored, the retainage (the amount withheld until final payment), the total amount of previous payments, and the amount of the current payment due.

2. You are the architect for a hotel project which has been completed and for which you have been paid in full. The owner informs you that he is so gratified with the plans and specifications that he intends to reuse them for an identical facility on a different site. What should be your response?
   A. Thank the owner and allow him to reuse the documents, since he legally has that right anyway.
   B. Bill the owner for 50 percent of the original fee, in accordance with the General Conditions.
   C. Submit the case for arbitration to determine whether the owner may reuse the documents.
   D. Inform the owner that the documents are your property and may not be used on any other project.

   The AIA General Conditions (AIA Document A201) are quite explicit on this point. Paragraph 1.3.1 states that "all drawings, specifications and copies thereof furnished by the architect are and shall remain his property. They are to be used only with respect to this project and are not to be used on any other project." D is therefore the correct answer.

3. How is the term "date of substantial completion" best defined?
   A. The date determined by the architect when the construction is sufficiently complete so the owner can occupy the building.
   B. The date when the building inspector makes his final inspection and certifies that the building may safely be occupied.
   C. The date when the liquidated damages clause in the construction contract takes effect.
   D. The date when 90 percent of the total contract price becomes due and payable.

   Close to the completion of a project, the contractor sends the architect a statement claiming substantial completion, at which time the architect makes a semi-final inspection. If the architect determines that the construction is sufficiently complete in accordance with the contract documents so the owner can occupy the project, he prepares a Certificate of Substantial Completion, to be signed by the contractor and the owner. The form often used for this purpose is AIA Document G 704. The correct answer is therefore A.

4. The bid results indicate that the low bidder for the construction of a science wing at a public college is approximately 25 percent lower than the next lowest bid. The apparently successful bidder reviews his calculations and sub-bids and informs the owner that he has made a number of substantial errors requiring him to withdraw his bid. Which of the following courses of action are open to the architect?

   I. Rejecting all bids, readvertising and rebidding the project.
   II. Recommending the low bidder forfeit the bid bond.
   III. Increasing the amount of the contract by a percentage of the amount of the low bidder's error.
   IV. Awarding the contract to the next lowest bidder.
   V. Negotiating a contract with the next lowest bidder for an amount less than his bid.

   A. I only
   B. III only
   C. II and IV only
   D. II and V only

It is unlikely that an architect would recommend forcing a successful bidder to enter into an agreement for the construction of a project with the knowledge that the contractor would probably lose money in the process. Furthermore, under those conditions, the contractor might attempt to cut corners in order to minimize his loss. The result: an unhappy contractor with no incentive to do a good job and an even unhappier owner with a potentially shoddy project. In this example, the difference between the low bid and the next lowest bid (25 percent) is substantial and indicates a legitimate error in calculation. The amount of the bid deposit or bid bond is usually between 5 percent and 10 percent of the architect's estimate or the amount of the bid proposal. Where the low bidder refuses to enter into a contract, the architect would, most likely, recommend that the amount of the bid security be forfeited to the owner to serve as liquidated damages but not as a penalty. Assuming that the next lowest bidder's proposal is still within the amount of available funds, the recommendation would be to award a contract for the next lowest bid. The forfeiture of the bid security reduces the amount of damages to the owner. The correct answer is C.

5. The bids for the construction of a public high school are tabulated as follows:

	Smith	French	Kline
Base Bid	$8,400,000	$8,100,000	$8,450,000
Additive Alternate A	410,000	490,000	310,000
Additive Alternate B	170,000	230,000	120,000
Additive Alternate C	220,000	420,000	170,000
Additive Alternate D	260,000	400,000	450,000

The total available funds for development, which includes a 5 percent construction contingency, equal $9,502,500. Which of the following actions should the architect recommend to the owner under these circumstances?

A. Reject all bids and readvertise deleting Alternate D since the sum of the base bids plus all alternates exceeds the available funds for construction.

B. Award the contract to French on the basis of his low base bid and negotiate change orders for the alternates after signing the contract.

C. Award the contract to Kline for the base bid plus Alternates A, B, and C since it is within the budget, the construction contingency is not affected, and Alternate D constitutes the lowest priority of work included in the project.

D. Award the contract to Smith for the base bid plus Alternates A, B, C, and D since the total is within the available funds for development.

Prior to analyzing the bid tabulation, we must determine the available funds for construction and the amount of the construction contingency. If the total budget for development is $9,502,500, the construction budget is $9,050,000 and the 5 percent contingency is equal to $452,500. The quickest way to calculate this is to divide the total by 1.05 (105 percent); i.e., $9,502,500 ÷ 1.05 = $9,050,000. Since the construction contingency must be held aside to allow for unforeseen additional costs during construction, the amount of the contract award may be for $9,050,000 or less. Additive alternates allow the owner to award a contract for the minimum acceptable amount of work under the project and offer a means to increase that award for increments of additional work up to and including the maximum cost possible within the available funds for construction. Since the sum of the three base bids plus all alternates exceeds the construction budget of $9,050,000, the contract must be awarded for the sum of the base bid plus those alternates that do not exceed the construction budget. Kline's bid, including Alternates A, B, and C, totals $9,050,000. The other two exceed that amount. Furthermore, additive alternates on public work are almost always accepted in their order of priority; i.e., A first, B second, and so forth, to avoid manipulating bids. For example, if we reject Alternate A and accept Alternates B, C, and D only, Smith becomes the low bidder and the total is within the budget for construction. If Smith were awarded the contract under those circumstances, Kline would likely enter into litigation. Choice A is not viable; as long as the base bid is within the available funds, a contract must be awarded. Awarding a contract to French on the basis of his low bid while rejecting all alternates may be legal but would probably result in increased costs to the owner, since change orders during construction are usually more costly than including those items in the original contract. Consequently, choice B is incorrect. Choice D is not correct since the sum of the base bid and all alternates is within the development budget, but would leave no additional funds for contingencies. The only possible recommendation is as stated in choice C.

6. The Performance Bond, AIA Document A311, stipulates that when a contractor is in default under the contract for construction, the surety must remedy the default. Which of the following courses of action may the surety take under the conditions of this obligation?

   I. Complete the contract in accordance with its terms and conditions.
   II. Obtain bids for completing the contract and arrange for a contract between the owner and low bidder for the remaining work.
   III. Pay for the cost of completing the work to the amount of the penal sum of the bond.
   IV. Employ the forces of the defaulted contractor to complete the work.

   A. I, II, III, IV
   B. I, II
   C. I, III
   D. III

The Performance Bond guarantees performance in accordance with the terms of the construction contract. Obligations under the bond are identical to those of the construction contract, and any inadequacies in the contract will constitute similar inadequacies in the bond. If the surety has to take over upon the contractor's default, it is obliged to perform the work in accordance with the contract documents to the amount of the penal sum of the bond. Since the surety has no liability for costs exceeding this amount, performance bonds are usually written to cover 100 percent of the contract price. In event of a default, the surety may proceed according to its own judgment as to the best way to complete the work. It may hire another contractor to do so, employ the forces of the defaulted contractor, or accomplish the uncompleted portion of the work with its own forces. Regardless of the procedure, the surety is liable for the cost to complete the work with no additional cost to the owner. All of the statements are correct, as itemized in answer A.

7. The purpose of the Labor and Material Payment Bond, AIA Document A311, is to

   I. prevent the potential filing of liens by workmen and material suppliers.
   II. assure faithful performance by the contractor and all subcontractors.
   III. protect the owner from claims made for unpaid contractors' bills after final payment has been made.
   IV. provide a means to discharge liens that are filed after the completion of the project.
   V. guarantee that the contractor will adhere to the requirements of the technical specifications.

   A. I, III   C. III, IV
   B. II, IV   D. III, V

A Labor and Material Payment Bond guarantees that the contractor's bills for labor and materials incurred under the contract are paid. The existence of this guarantee protects the owner from liens and other claims made after the completion of the project and after final payment has been made to the contractor. This does not, however, guarantee that liens will not be filed against the project, choice I. Anyone furnishing labor or materials may file a lien against the project. The bond, however, protects the owner from potential liabilities resulting from such liens being filed and the surety is obligated to discharge such liens through payment of all unpaid bills. The faithful performance, choice II, as well as the adherence to specifications, choice V, is guaranteed by the Performance Bond. Choices III and IV correctly describe the purpose of the Labor and Material Payment Bond; the correct answer is therefore C.

8. Which of the following would not be considered to be a part of the Construction Contract Documents?
   I. General and Supplementary Conditions
   II. Invitation to Bid
   III. Instructions to Bidders
   IV. Addenda
   V. Change Orders
   A. I, III, IV
   B. II, III, V
   C. II, III
   D. III, IV

This question tests a condidate's basic knowledge of the general conditions of the contract and the definition of contract documents. Furthermore it is a negatively worded question that asks that you identify those items not normally considered to be a part of the contract documents. Questions of this type require the candidate's familiarity with relevant AIA documents, in this case the General Conditions of the Contract for Construction, AIA Document A201. By reversing the process of elimination, we can quickly determine which of the items suggest contractual implications in terms of the project's cost, intent, and execution. It is apparent that I, IV, and V fall into that category enabling us to eliminate choices A, B, and D. Consequently, we can select answer C with the understanding that both the invitation to bid and the instructions are merely for the convenience and information of bidders and do not constitute any contractual implications related to the project's construction.

9. The Base Bid is the amount stated in the bid as the sum for which the bidder offers to perform the work
   A. before any alternates are considered.
   B. including all deductive alternates.
   C. including all additive alternates.
   D. not including any cash allowances.

The Base Bid usually constitutes the amount of the contract sum proposed by the bidder for the completion of work defined in the Base Bid Documents. Alternates, either additive or deductive, are variations in contract requirements described in the Specifications and indicated on the drawings on which a separate price is received as part of the bid. The owner normally reserves the right to accept none, any, or all alternates, depending on the amount of the base bid and its relation to the project budget. Cash allowances are established in the contract documents for inclusion in the contract sum to cover the cost of prescribed items not specified in detail. Variations between such amounts and the ultimate cost of these prescribed items are reflected in change orders adjusting the contract sum. The correct answer is A.

10. Which of the following subjects are correctly matched with the document that must contain an accurate and detailed enumeration of these subjects?
    I. Safety
    II. Surveys and Permits
    III. Addenda
    IV. Time of Substantial Completion
    V. Final Payment
    VI. Specialties
    A. General Conditions—I, IV, VI
    B. Special Conditions—I, II, V
    C. Technical Specifications—II, III, VI
    D. Owner-Contractor Agreement— III, IV, V

This is an example of one of the more difficult types of questions, because it calls for a considerable amount of mental gymnastics in analyzing the answer. Using the process of elimination is probably the best way to discover the correct answer; consequently, we can examine the four choices and eliminate those that are incorrect. A knowledge of the content and purpose of each document is essential in this process. The General Conditions (A), make reference to Safety of Persons & Property and Safety Precautions & Programs in general terms. More detailed requirements are spelled out in the Special Conditions, and they are occasionally referenced in the Specifications. Final Completion and Final Payment are also covered in the General Conditions, without referring to a specific date. Specialties, however, is a term reserved for one of the headings in the Technical Specifications and describes items such as chalkboards and tackboards, metal toilet partitions, flagpoles, identifying devices and so forth. Answer A can be eliminated for that reason.

None of the subjects listed, other than Safety, are included in the Special Conditions. Special Conditions are usually conditions peculiar to the project that must be mentioned, in addition to the General Conditions and the Supplementary General Conditions—the latter being for the purpose of modifying the General Conditions. Therefore answer B can be eliminated. Specialties, as discussed above, is the only subject listed that appears in the Technical Specifications. All other subjects are conditions to the contract and are never included in this document. Answer C can be eliminated. The Owner-Contractor Agreement, also known as the contract for construction, specifies the time of completion of construction and when and under what conditions final payment is made. Furthermore, this agreement itemizes all contract documents, i.e., drawings, specifications, etc. The Addenda are listed in the enumeration of contract documents in order to assure that the work included in the contract is based on the latest information submitted to all bidders of the project. Consequently, answer D is correct.

11. Select the statement which most closely describes the purpose of state lien laws.
    A. They assure payment of all bills owed by the general contractor.
    B. They provide a means of filing a claim against the owner and his property for personal injury to workmen.
    C. They are for the express purpose of insuring successful completion of the project should the contractor or subcontractors default during construction.
    D. They prevent the owner from withholding payment of the contractor's retainage for more than 35 days after substantial completion.

Some of the choices in this type of question contain half-truths that can be misleading unless one understands the purpose of state lien laws. Lien laws assure satisfactory payment to workmen and suppliers of all debts incurred on behalf of the project. The owner of the property is liable for such payment to those who supply materials for or perform work on the project. The purpose of these laws is to protect persons performing work or providing materials by affording them security for their just compensation. The General Conditions stipulate that neither the final payment nor any portion of the retainage is to be paid until the contractor provides a release or waiver of liens to the owner. It is essential that the owner receive this protection against liens so that he is free to dispose of the property with no encumbrances to preclude a transfer of title. The correct answer is A.

DOORS, WINDOWS, & GLASS

1. Glass is made in many types suitable for a variety of applications in building construction. Match each type of glass with the correct description or application.

   I. Heat-strengthened glass
   II. Laminated glass
   III. Plate glass
   IV. Structural glass
   V. Tempered glass

   1. Safety glass
   2. Ground and polished to a flat plane
   3. Impact resistance
   4. Spandrel glazing in curtain wall systems
   5. Facing material for vertical or horizontal surfaces

   A. I-4, II-1, III-2, IV-5, V-3
   B. I-5, II-3, III-2, IV-1, V-4
   C. I-3, II-1, III-5, IV-2, V-4
   D. I-1, II-3, III-5, IV-4, V-2

Heat-strengthened glass has a colored ceramic glaze fused to it to make it opaque; it is used principally for spandrel glazing in curtain wall construction, and its strength is about twice that of ordinary glass (I-4). Laminated glass, also called safety glass (II-1), consists of two or more layers of glass bonded to a transparent plastic sheet sandwiched between them to form a shatter-resistant assembly. Plate glass is glass that has been ground and polished on both sides to a perfectly flat plane (III-2). Structural glass is a special opaque, colored glass used as a facing material for vertical or horizontal surfaces of walls and partitions (IV-5). Finally, tempered glass is glass that has been reheated and then suddenly cooled to produce a glass two to four times stronger than ordinary glass in resisting impact forces (V-3). Answer A correctly matches the terms and descriptions.

2.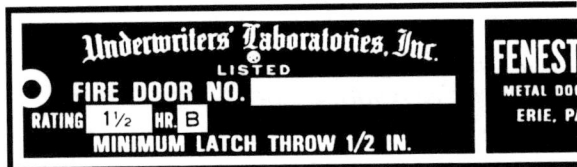

   If the label shown above were attached to a door, it would indicate that

   I. the door requires an automatic closing device.
   II. the fire assembly is self-closing.
   III. no glazed openings are permitted in the door.
   IV. 100 square inches of glazed openings are permitted in the door.

   A. I and III   C. I and IV
   B. II and IV   D. II and III

Openings in walls with fire-resistive ratings are classified in accordance with the character and location of the wall in which they are located. Class A openings are in walls separating buildings or dividing a single building into fire areas. Doors in these openings must have a fire-protection rating of three hours, may not have any glazed openings, and must be of an automatic closing type. That is, the door remains in an open position but will close automatically if subject to an increase in temperature. Statements I and III therefore apply to Class A openings. Class B openings are in enclosures of vertical communication through buildings, such as stairs and elevators. Doors in Class B openings must have a rating of one or 1 1/2 hours, may have not more than 100 square inches of glazed openings, and must be self-closing. In other words, the door is normally closed and is equipped with a device to insure closing and latching after having been opened for use. Thus, II and IV are applicable to Class B openings. The door label shown designates a 1 1/2 hour fire-resistive rating used in Class B openings. Since II and IV are correct, B is the right answer.

3. You are standing in a corridor facing a door with the knob on the left hand side. Since the door opens into the room, it must be a
   A. right hand door.
   B. left hand door.
   C. right hand reverse door.
   D. left hand reverse door.

In selecting door hardware, it is important to know the type of door swing and the direction of swing. For this purpose, you must imagine that you are standing facing the outside of the door. The location of the hinges determines the hand of the door: if the hinges are on the left hand side, it is a left hand door, and if the hinges are on the right hand side, it is a right hand door. If the door opens away from you, into the room, then it is a regular swing door; and if the door opens towards you, to the outside, it is a reverse swing door. The outside of the door is defined as any one of the following: (1) The street side of an entrance door. (2) The corridor side of a room door. (3) The side opposite the hinges for a door between rooms. In this case, you are outside a door with the knob on the left and therefore the hinges on the right (right hand door). Since the door swings away from you, it is a regular swing door. The only correct answer, therefore, is A.

4. Select the *incorrect* statement.
   A. Kalamein doors consist of a solid wood core covered with sheet metal, and can resist fire as long as the sheet metal cover prevents oxygen from reaching the core.
   B. Flush doors always have a core of solid softwood made of individual core blocks glued together.
   C. Although revolving doors are frequently used at entrances to major buildings, they cannot be used as legal exits.
   D. The glass in a pivoted glass door cannot be cut or drilled after fabrication.

Kalamein doors, also called metal clad doors, are correctly described in statement A. In recent years, kalamein doors have largely been replaced by fire-rated solid-core doors. Flush doors (B) may be made either with a solid core, as described, or with a hollow core—that is, a core made of small pieces of wood arranged in a grid pattern. B is therefore incorrect. Statement C is correct; because they can carry a continuous two-way flow of pedestrian traffic without much interchange of air between inside and outside, revolving doors are often used at entrances to large buildings. However, the UBC does not permit revolving, sliding, or overhead doors to be used as legal exits. Glass doors (D) are generally pivoted at the top and bottom, and are made of tempered glass, which cannot be cut, drilled, or altered in any way. D is therefore correct. The only incorrect statement, and therefore the answer to this question, is B.

5. Which of the following statements about wood windows is true?

   A. They are usually more expensive than steel or aluminum windows.

   B. They are manufactured from kiln-dried hardwood.

   C. They are required to be treated against fungi, insects, and water.

   D. They are usually made at the site to fit the actual opening dimensions.

Wood windows are fabricated in a mill (D) from kiln-dried softwoods (B) and treated to resist moisture, mold, fungi, and insects (C). Though generally less expensive than metal windows (A), wood windows are durable and available in a variety of types and standard sizes. The only true statement is C.

EARTHQUAKE

1. Select the correct statement.
   A. Earthquakes cause horizontal vibrations only.
   B. Earthquakes cause vertical ground vibrations only.
   C. Earthquakes cause both horizontal and vertical ground vibrations, but the vertical vibration is generally not considered in design.
   D. Earthquakes cause both horizontal and vertical ground vibrations, but the horizontal vibration is generally not considered in design.

   An earthquake causes the ground to shake both horizontally and vertically. The vertical motions are generally neglected in design because they are usually smaller than the horizontal, and also because of the considerably greater stiffness of buildings in the vertical direction. C is the correct answer.

2. In the seismic formula V = ZIKCSW, the value of C depends on
   A. the location of the building.
   B. the height and length of the building.
   C. the building's dead load.
   D. the type of structural system used.

   The total horizontal seismic force V acting on a building is equal to the product of six factors: Z, I, K, C, S, and W, where
   Z is the zone coefficient (choice A).
   I is the occupancy importance factor
   K is a factor depending on the type of structural system used (choice D).
   C is a coefficient related to the flexibility of the structure.
   C is a function of the fundamental period of vibration, which in turn depends on the height and plan dimensions of the building (correct choice B).
   S is a factor accounting for site-structure resonance.
   W is the total dead load of the building (choice C).

3. Four causes of earthquake damage to buildings are listed below. The earthquake regulations of the Uniform Building Code are intended to provide resistance to which of them?
   I. Ground rupture in fault zones.
   II. Ground failure.
   III. Tsunamis.
   IV. Ground shaking.
   A. All of them
   B. II and IV
   C. IV only
   D. I and II

   The earthquake regulations of the UBC are intended to provide resistance to ground shaking (IV), as noted in correct answer C. The code does not provide for resistance to ground ruptures (I); ground failure (II), such as landslides, subsidence, or settlement; or tsunamis (III), which are seismic seawaves.

4. Adjacent buildings should be sufficiently separated to prevent them from pounding one another during an earthquake. In this regard, which of the following is the most correct statement?
   A. The separation should be extended through the foundation.
   B. The drift value of each building should be calculated, and the separation should be at least equal to the greater drift.
   C. The amount of separation depends on the height of the building, and should be at least one-half inch for each 20 feet of height.
   D. The separation should at least equal the sum of the calculated drift values of each building.

During an earthquake, every building moves differently. The amount of horizontal movement of a building from its original vertical position is called drift. Since adjacent buildings can move out of phase of each other during an earthquake, the separation between them should at least equal the sum of the calculated drift values of each building (correct answer D), or the two buildings will hammer against each other during an earthquake, causing considerable damage. Let's look at the other three choices. A is incorrect because joints or separations generally extend to the top of the foundation, not through it. The separation as described in choice B is insufficient, as explained above. Finally, the first part of C is correct— the amount of separation does vary with the height of the building, being greater for taller buildings. However, the stated separation of 1/2 inch for every 20 feet of height is much too low. Older editions of the UBC required the separation to be at least one inch plus 1/2 inch for every ten feet of height above 20 feet.

5. Four building plan shapes are shown. Which is best able to withstand earthquakes?

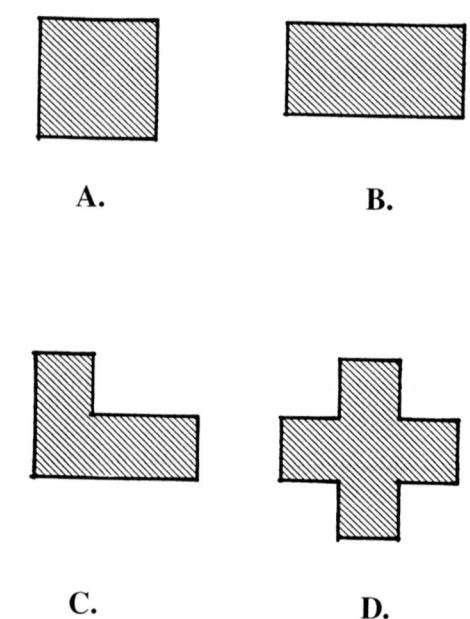

A. 　　　　B.

C. 　　　　D.

In buildings with an irregular shape, such as L, T, or cruciform configurations (C and D), the less rigid wings tend to rotate about the more rigid elements during an earthquake. This torsional movement can result in severe damage, particularly where two wings join. That leaves us with choices A and B. The best basic plan shape is one which is symmetrical and equally capable of resisting earthquake forces imposed from any direction. The square plan (correct answer A) is therefore preferable to the rectangular plan (B).

6. A building in seismic zone 4 has a ductile moment-resisting space frame capable of resisting the total required lateral force. Its total dead weight is 10,000 kips. I and S are each equal to 1.0. Its fundamental period of vibration is 0.5 seconds. What is the total lateral force from earthquake?

   A. 804 kips
   C. 752 kips
   B. 630 kips
   D. 335 kips

   The basic seismic formula is
   V (total lateral force) = ZIKCSW
   In this case, Z = 1 for seismic zone 4
   I = 1.0, K = 0.67, in accordance with Table 23-I of the Uniform Building Code

   $C = 1/15\sqrt{T} = 1/15\sqrt{0.5} = 0.094$
   S = 1.0
   W = 10,000 kips
   $V = 1 \times 1.0 \times 0.67 \times 0.094 \times 1.0 \times 10,000$
   = 630 kips (correct answer B)

7. What is the maximum allowable height of the building in Question 6?

   A. 120 feet
   C. 240 feet
   B. 160 feet
   D. No limit

   The only height restriction imposed by the Uniform Building Code is that buildings over 160 feet in height must have ductile moment-resisting space frames capable of resisting not less than 25 percent of the required seismic forces for the structure as a whole. Since the building has such a frame, its height is not limited by the code, making D the correct answer.

8. What is the purpose of a diaphragm in earthquake-resistant construction?

   A. To distribute horizontal forces to the vertical resisting elements.
   B. To provide mass in order to damp out vibration.
   C. To provide ductility so the structure can absorb energy in the inelastic range.
   D. To rigidly connect dissimilar building elements.

   A diaphragm is the horizontal floor or roof system that distributes horizontal forces to the vertical resisting elements, such as shear walls or moment-resisting frames (correct answer A). Among the diaphragm materials and systems in common use are reinforced concrete slabs, steel decking, and plywood sheathing.

ELECTRICITY & LIGHTING

1. In a 120 volt direct current lighting system, the correct procedure for obtaining a 12 volt power outlet is to provide a

    A. 10 KVA transformer.
    B. 120v/12v transformer.
    C. resistor in parallel with the 12 volt power outlet.
    D. resistor in series with the 12 volt power outlet.

    Since transformers do not operate on direct current, both choices A and B are incorrect. Placing a resistor in parallel with the 12 volt outlet (C) would not change the voltage. However, placing a resistor in series with the 12 volt outlet would create the necessary voltage drop. D, therefore, is the correct answer.

2. A transformer having 3,000 turns on the primary winding and 600 turns on the secondary winding is supplied with 120 volts A.C. What will the secondary voltage be?

    A. 12 volts C. 48 volts
    B. 24 volts D. 96 volts

    A transformer is a device that either increases or decreases the voltage of an AC circuit, generally consisting of two windings, a primary and a secondary. The number of turns of wire on each winding is proportional to the voltage rating of the winding; that is, the winding having more turns will have the higher voltage and the winding having fewer turns will have the lower voltage. In this case, the primary winding has 3,000 turns and is supplied with 120 volts, while the secondary winding has 600 turns. Since the voltage is proportional to the number of turns,

    $$\frac{primary\ voltage}{primary\ turns} = \frac{secondary\ voltage}{secondary\ turns}$$

    $$\frac{120}{3000} = \frac{v_2}{600} \qquad v_2 = \frac{600}{3000} \times 120 =$$

    24 volts (correct answer B)

3. Three-phase, four-wire electrical systems in modern buildings serve

    A. lighting.
    B. power.
    C. lighting and power.
    D. motors.

    A three-phase electrical system has at least three wires and is identified by the voltage between any two of the wires, 480 volts being common. Sometimes a fourth wire, the neutral wire, is provided, which is normally grounded at the source for safety. The voltage between the neutral and any other wire is equal to the phase-to-phase voltage divided by 1.73. For example, where the phase-to-phase voltage is 480 volts, the phase-to-neutral voltage is 480 ÷ 1.73 = 277 volts. This system is then referred to as 480/277 volt three-phase power. The three-phase, four-wire system is very versatile; three-phase loads, such as motors and transformers, can be connected to the three-phase wires; and single-phase loads can be connected between any two-phase wires, or between any phase wire and the neutral wire, depending on the voltage rating of the load. Such systems can therefore serve all lighting and power needs, including motors. C is therefore the best answer.

4. Disconnect protection for motor loads is often set higher than the conductor size rating for which of the following reasons?
   A. To permit the motor to start.
   B. To protect the motor when it is running.
   C. To protect the motor against overload.
   D. To allow for the inaccuracy of the overload device.

The instant a motor starts, an inrush current flows, which is larger than the current that flows during normal operation. The wiring to the motor can readily conduct this larger current flow because of its thermal mass. However, if the disconnect protection were set exactly at the current rating of the wiring, it would quickly open and break the circuit, since it is a very sensitive device that responds quickly to even a slight rise in temperature. Therefore, such protection is often set higher than the conductor size rating (correct answer A).

5. In an overhead power district, the power company may provide service
   I. underground.
   II. overhead.
   III. underground if the owner provides the conduit between the building and the power pole.
   IV. underground if the conduit is encased in concrete.
   A. II only
   B. I, II, III, IV
   C. II and III
   D. III and IV

In an underground power district, the power company must provide service running underground to buildings. In overhead districts, they only need to provide service overhead. However, if a building owner in an overhead district desires underground service, the power company may elect to give him such service if the owner provides the underground conduit between the building and the power pole. The conduit needs to be encased in concrete only if the power is at a very high voltage. Therefore C is the correct answer.

6. The number of lumens produced by a light source
   A. varies directly as the distance between the source and the lighted surface.
   B. varies inversely as the distance between the source and the lighted surface.
   C. varies inversely as the square of the distance between the source and the lighted surface.
   D. is independent of the distance between the source and the lighted surface.

Candidates should understand the difference between the light produced by a lamp and the lighting level on a surface. Lamps are generally rated by the amount of light they emit, in lumens. The lighting level on a given plane, however, is stated in foot candles, where one foot candle is equal to one lumen per square foot. The number of lumens produced by a lamp is a function only of the lamp and is independent of the distance to the work surface. D is therefore the correct answer.

7. Light reflectors and refractors are designed
   A. to provide a specific distribution of light.
   B. to reduce glare.
   C. to reduce veiling reflections.
   D. to increase visibility.

Reflectors and refractors provide a specific distribution of light; that is, direct, indirect, focused, diffused, etc. Although reduced glare and veiling reflections (reflected glare) may also be achieved with reflectors and refractors, that is not their primary purpose, and usually, the placement of the light source has a greater effect in this regard anyway. Increased visibility is generally accomplished by using additional light. Therefore A is the best answer.

8. Which of the following light sources provides the best color rendition properties for a beauty salon?
   A. High pressure sodium
   B. Low pressure sodium
   C. Mercury vapor
   D. Incandescent

In this question we should assume that the most desirable lighting for a beauty salon is the type that produces a warm, natural, and flattering quality of light. High pressure sodium and low pressure sodium (A and B) both have a very strong yellow spectrum. Although improvements in the color rendition of mercury vapor lamps (C) have been made, they still tend toward the blue end of the spectrum. Incandescent light (D), while not nearly as efficient a light source as the others, has the most continuous spectrum of light, and therefore the most natural appearing light. Incandescent lamps are therefore the best light source for a beauty salon (correct answer D).

9. Which type of lighting fixture provides the most diffuse light?
   A. Direct
   B. Indirect
   C. Semi-direct
   D. Semi-indirect

Lighting fixtures have a variety of light distribution patterns. Direct lighting fixtures (A) produce light which is undiffused and often glary. Indirect fixtures (B) produce highly diffuse illumination, since the light is reflected off a surface that scatters it so that there is virtually no direct light. Semi-direct (C) and semi-indirect (D) fixtures have distribution patterns in between. B is the best answer to this question.

10. Introducing controlled daylight into a space
    I. will increase glare.
    II. will increase visibility.
    III. will add solar heat gain.
    IV. can reduce the electric lighting load.
    A. I, II, III, IV
    B. II, III, IV
    C. III and IV
    D. I and II

Controlled daylight implies that glare and brightness will be minimized (I). Obviously the visibility will be increased (II) if additional light is added to a space. Since daylight is solar energy, it will produce solar heat gain (III). And because the visibility in the space is increased, it is possible to reduce the electric lighting (IV). This in turn will reduce internal heat gain, as well as reduce the energy consumed by the electric lights. Therefore B is the answer.

ENERGY CONSERVATION

1. How many BTU are required to raise the temperature of 100 gallons of water from 40°F to 150°F (4°C to 66°C)? Note that the weight of 100 gallons of water is 835 pounds, and its volume is 13.4 cubic feet.

   A. 1,474
   B. 11,000
   C. 51,770
   D. 91,850

With energy conservation becoming an increasingly important concern in the design of buildings, it is important for candidates to be familiar with basic terminology. A British thermal unit, abbreviated BTU, is defined as the amount of heat required to raise the temperature of one pound of water 1°F. Therefore, to raise the temperature of 835 pounds of water 110°F (from 40° to 150°), the number of BTU required is 835 × 110 = 91,850 (correct answer D). Candidates should be aware that some exam questions contain superfluous information. We have purposely included some unnecessary data here, such as the Celsius temperatures and the volume of water, in order to simulate such questions.

2. You are the architect for a 100,000 square foot office building, and the owner has asked you to incorporate energy-conserving techniques into your design. For purposes of analysis, you have separated the building's energy-consuming components into six categories as follows:

   I. Heating
   II. Cooling
   III. Domestic Hot Water
   IV. Vertical Transportation
   V. Lighting
   VI. HVAC Fans

   Rank these categories according to their average energy consumption, from highest to lowest.

   A. I, II, V, VI, III, IV
   B. II, I, VI, V, IV, III
   C. V, I, II, VI, IV, III
   D. V, VI, II, I, III, IV

In a recent study by the AIA Research Corporation, it was determined that lighting (V) is typically the largest energy user in an office building, accounting for about 35 percent of total building energy consumption. The other energy consumers, in order, are heating (I) (26 percent), cooling (II) (17 percent), HVAC fans (VI) (11 percent), vertical transportation (IV) (8 percent), and domestic hot water (III) (4 percent). The ranking is shown correctly in choice C.

3. Following are four statements concerning energy conservation. Which of them is true?

   A. In general, it is more energy-efficient to demolish an old building and replace it with a new one than to rehabilitate it.
   B. Since most buildings are unoccupied most of the time, significant savings during the heating season can be made by lowering the inside temperature during nights and weekends.
   C. Since corridors receive intensive use, their temperature during the heating season should be set higher than that in other areas.
   D. In order to conserve heating energy, the amount of cold outside air introduced into a building for ventilation should be limited to the amount required to replace expended oxygen.

Supporters of preservation have long contended that saving and restoring old buildings takes less energy than demolishing and replacing them. That idea has generally been proven to be correct (A is false) for two reasons. First, many old buildings actually require less energy for their operation than newer buildings. In addition, even when a new building is more energy-efficient in its operation, the cost in energy to tear down an old building and then construct a new building to take its place is often a more important factor in energy conservation than the energy to operate the building. Statement B is true and therefore the answer to this question. Considering nights, weekends, and holidays, most buildings are, indeed, unoccupied most of the time. Therefore, significant savings in heating energy can be made by lowering the temperature level during those times when a building is unoccupied. Corridors are actually unoccupied areas used mainly by people walking from one heated space to another. Corridor temperatures can therefore be set lower than temperatures in other occupied areas (false statement C). Although it is true that the amount of cold outside ventilation air should be limited in order to conserve heating energy, the amount required to replace expended oxygen is entirely inadequate. The amount of outside air introduced should be sufficient to get rid of unpleasant odors and smoke, as well. D is therefore false.

4. Select the correct statements about energy conservation from those that follow.
   I. Since the north and west sides of a building are often subject to cold winds, entrances on these sides should be avoided if possible.
   II. A square building will usually experience less heat gain or loss than a rectangular building of equal area.
   III. Heat transmission through glass can be reduced to a level comparable to that of a solid wall by using double- or triple-glazing.
   IV. Reducing the lighting level in a building also reduces the amount of heating and cooling required for the building.

   A. All of the above
   B. II, III
   C. I, II
   D. I, III

Energy conservation in buildings involves a great many variables, including the building's orientation, site, materials, fenestration, mechanical and electrical systems, etc. There is no single approach to energy conservation; rather, it is important to have an understanding of all the various factors that affect a building's use of energy. Although a site's microclimate must be considered, it is generally true that the north and west sides of a building are most subject to cold winds (I). If possible, therefore, entrances and glazed openings on these sides should be avoided. The heat gain or loss of a building is a function of its surface area. Since a square building has less surface area than a rectangular building of equal floor area, it experiences less heat gain or loss (II). Double-glazing has a U value of about 0.60, about half that of single-glazing. Triple-glazing has a U value of about 0.40. But the U value of a solid wall is even lower—an insulated masonry or concrete wall has a U value of around 0.10 to 0.20, which can be reduced to 0.04 or even less. III is therefore untrue. A building's lighting system also produces heat. Reducing the lighting level, therefore, has two effects: in the summer, with less heat given off by the lighting, less cooling for the building is needed. During the heating season, however, with the lighting system generating less heat, more heat must come from the building's regular heating system. So, reducing the lighting level in a building increases *the amount of heating required and* decreases *the amount of cooling required for the building. IV is therefore incorrect. Answer C is correct.*

5. Select the *incorrect* statement.

   A. The most effective sun control for the south wall of a building is a horizontal overhang.

   B. Since the east and west walls of a building receive more solar radiation in winter than in summer, fenestration in these walls is desirable to maximize heat gain in winter.

   C. The use of reflective and heat-absorbing glass is most effective in the summer.

   D. Heavy walls are most effective in reducing energy consumption where diurnal temperature variations are greatest.

In the Northern Hemisphere, the sun's path is inclined southerly and is higher in the summer than in the winter. Consequently, horizontal overhangs for south walls block the high summer sun, while allowing the low winter sun to enter (correct statement A). Statement B is incorrect and therefore the answer to this question. East and west facades receive more solar radiation in summer than in winter, and therefore minimum fenestration in these walls is desirable in order to avoid excessive summer heat gain. C is correct; most of the sun's radiant energy can be prevented from entering a building by using reflective or heat-absorbing glass. However, use of these special types of glass also reduces the amount of natural light and blocks the warmth of the winter sun. D is also a correct statement; heavy materials store peak heat loads and later release them when the outside temperature drops. In areas having high diurnal (daily) temperature variation, the use of heavy walls extends the coolness of night to the day and the heat of the day to the nighttime, which results in a saving of energy.

FINANCING

1. Municipalities often issue various types of bonds to pay for capital improvements. In this connection, select the correct statements.

   I. Only general obligation bonds are exempt from federal income taxes.

   II. General obligation bonds are backed by the full faith and credit of the government.

   III. Revenue bonds are secured by the revenue from a self-liquidating project.

   IV. Mortgage bonds are backed by a mortgage on a utility or other property.

   V. A statutory or constitutional tax limit is invalid when such limit must be exceeded in order to pay the principal and interest on bonds.

   A. I, IV
   B. II, III
   C. II, III, IV
   D. All of the above

   Bonds are issued by various governmental entities, including states, counties, cities, special districts, etc. The three most common types of bonds are general obligation, revenue, and mortgage. General obligation bonds are backed by the full faith and credit of the governmental unit (true statement II). Revenue bonds are secured by the revenue from a self-liquidating project, such as a toll bridge, tunnel, or parking garage (true statement III). Mortgage bonds are generally used in connection with the purchase or construction of utilities, and they are secured by a mortgage on the utility (correct statement IV). I is not a true statement; the income from all municipal bonds, not just general obligation bonds, is exempt from federal income taxes. V is also false; sometimes there is a statutory or constitutional limit to the tax-levying authority of a municipality. Such a limit remains in effect unless the law is changed or the constitution amended. The fact that the municipality has certain obligations to meet does not invalidate the tax limit. The correct combination of true statements appears in answer C.

2. Select the *incorrect* statement about financing from those that follow.

   A. Takeout financing is the same as permanent financing.

   B. A construction loan is used for interim financing.

   C. In computing equity return, a construction project cannot be assumed to appreciate and depreciate at the same time.

   D. A project with a negative cash flow can show a total rate of return that is positive.

   As the only generalist in the construction process, an architect is expected to have general knowledge about real estate financing. Let's examine each of the statements above. A loan that takes over a short-term construction loan on the basis of a permanent mortgage is called takeout financing (true statement A). B is also true; interim financing is short-term, temporary financing generally in effect during the construction of a project. Although it sounds paradoxical, equity return computations can assume that a project appreciates and also depreciates, making C an incorrect statement and therefore the answer to this question. To explain further: one of the factors considered in the financial analysis of a construction project is the income tax that the owner must pay (or the tax benefit which he will derive) from the taxable income (or loss) from the project. Taxable income (or loss), in turn, is partially a function of the assumed depreciation of the project. Even when it is believed that a project will actually increase in value, depreciation can be considered an expense for income tax purposes. Yet in calculating the total rate of return, the expected appreciation of the property is considered. Hence, both depreciation and appreciation, though they are opposites, can be assumed to occur. Finally, D is true. Total rate of return takes into account not only cash flow, but also principal build-up as the loan is paid off, and appreciation of the property. If these latter two factors are greater than the negative cash flow, the result will be a total rate of return that is positive.*

3. That portion of a construction loan that must be paid to the lender for the privilege of borrowing the money is called the

   A. discount.
   B. recording fee.
   C. balloon payment.
   D. amortization

   A construction loan is short-term interim financing, which is in effect during the construction of a project. For smaller projects, construction loans are often made through banks or savings and loan associations rather than mortgage brokers, because this procedure is usually less costly and time consuming. Regardless of who the lender is, the borrower must pay certain fees at the time he obtains the loan, including the discount, legal fees, and recording fees. The discount (A) may be known by various names—loan placement charge, origination fee, "points"—and represents a fee charged by the lender to the borrower for the privilege of obtaining the loan. For example, if the lender states that the discount will be two points, that means that for every $1,000 borrowed, the borrower will actually receive only $980, the difference of $20 being two points (i.e., 2 percent) of the $1,000 borrowed. But the borrower will have to repay the full $1,000, plus the interest on $1,000. In effect, this raises the true interest rate paid by the borrower. The lender's justification for the origination fee is that it is required to pay the costs of the loan officer as well as other loan processing costs. Recording fees (B) include charges made by the county to record the mortgage, as well as any taxes levied by the state and federal government on the transaction. A balloon payment (C) is a final mortgage payment that is much larger than the typical periodic payments. And finally amortization (D) refers to the payment of a debt in periodic installments. The correct answer is A.

4. The cash derived from refinancing a property that has been held for less than five years

   A. is tax free.
   B. is taxable.
   C. reduces the size of the existing loan.
   D. must be reinvested within 18 months to avoid capital gains taxes.

   Part of each mortgage payment is interest and the rest is applied to the principal. Consequently, the amount of indebtedness, or unpaid principal, is reduced with each payment. As the years go by, with the property having less indebtedness against it and possibly having appreciated in value, it may be desirable to refinance the property, that is, to increase the amount of the loan, in order to obtain cash. The cash derived from refinancing is not taxable (correct answer A), simply because it is money that is borrowed and not derived from either earnings or profit. The five-year period of time mentioned in this question is totally irrelevant.

5. A 25-year bond issue is used to finance the construction of a $26 million basketball arena at a state university. Assuming an annual interest rate of 7.5 percent, the total debt service would be approximately

   A. $2,210,000.
   B. $24,375,000.
   C. $48,750,000.
   D. $55,250,000.

   It is common for public agencies to issue general obligation or revenue bonds in order to finance capital improvements. The advantages of using such funding methods are twofold: first, the project can proceed when it is needed, instead of having to wait for sufficient tax money to be available, and secondly, repayment of the debt is made over a long period of time with potentially cheaper money. The total debt service of a bond issue is the sum of the bond issue amount and the interest over the total financing period. In this case, it is the sum of $26 million and the total interest costs for the 25-year period. A good rule of thumb for estimating total interest costs is to multiply the amount of the bond issue by the interest rate for one year ($26 million times 7.5 percent), multiply that result by the number of years until the bonds mature, and multiply that result by 60 percent, since the annual interest declines as the debt declines. Therefore, $26,000,000 × 0.075 × 25 × 0.60 = $29,250,000. Since the interest must be added to the principal, the total debt service in this case is $26,000,000 + 29,250,000 = $55,250,000. The correct answer is D.

6. A "default in prior mortgage" clause is commonly inserted in

   A. first mortgages.
   B. second mortgages.
   C. reduction certificates.
   D. property deeds.

   When a buyer of property is unable to finance the difference between the cost of the property and the amount of the first mortgage, he often obtains a second mortgage. The second mortgage is subordinate to the first and usually carries a higher interest rate than the first because of the greater risk assumed by the second lender. A "default in prior mortgage" clause is commonly included in second mortgages (correct answer B), so that if the mortgagor (borrower) defaults in payment on the first mortgage, the second mortgagee (lender) may pay the amount, add it to his loan, and institute a foreclosure action. A reduction certificate (C) is a document showing the balance due on a mortgage at the time of the sale of the property.

7. A sale-and-leaseback agreement affords the owner an opportunity to

   A. increase his potential profit on the sale of his newly-completed project.
   B. minimize his capital investment.
   C. realize tax benefits from depreciation of his property not otherwise possible.
   D. get out of a bad investment.

   In a sale-and-leaseback transaction, the owner-developer of a project is able to occupy the property under a long-term lease while, at the same time, recovering his capital investment through the sale of the project to another owner. His lease payments become a tax-deductible business expense although, since he no longer owns the property, he cannot continue to depreciate his investment for the purpose of sheltering income. The correct answer is B.

8. A $100 per student per semester fee is the principal source of revenue to repay a proposed bond issue for the planned construction of a new student union to be located on a state college campus. Based on the college's preliminary assessment of space needs, the project cost will exceed the proposed bond issue by approximately 25 percent. In response to the owner's request for guidance, the architect would most likely recommend which of the following?

   A. Increase the student fee to $125 per semester and proceed with the project.
   B. Delay the project until the student enrollment has increased to the point where it can support the larger bond issue required for the project.
   C. Proceed with the project on the basis of the available funds and attempt to reduce the scope or quality of work during the programming phase.
   D. Request the owner to reduce his gross space needs sufficiently to meet the budgetary constraints prior to entering into an agreement for professional services.

This type of question tests the candidate's ability to use his judgment in situations where financing dictates cost, scope, and quality of a project. It would be presumptuous on the architect's part to recommend a 25 percent increase in student fees (A). Such an action would require the approval of the college's administration and student body—a policy matter with which the architect should not concern himself. Delaying the project (B) would almost certainly result in an escalation of cost and an increased budget. Furthermore, the architect is not qualified to project future student enrollments and would, once again, be acting outside his area of expertise. Proceeding with a project for which adequate funds are not on hand (C) is unrealistic. The architect cannot arbitrarily reduce space needs 25 percent, nor can he reasonably expect to reduce the project's quality 25 percent. The only reasonable course of action is to convince the owner to reduce his overall space needs to meet the budget prior to entering into an agreement (D), and to attempt to develop efficient uses of space that will satisfy the users' functional needs. The correct answer is D.

FINISHES

1. Which of the qualities listed below are among the unique properties that make weathering steel a desirable exterior material?

   I. It does not rust.
   II. It requires no protective coating.
   III. It is virtually maintenance free.
   IV. It ultimately weathers to a deep brownish-red finish that resists corrosion.
   V. It provides an excellent base for exterior finishes.

   A. I, II, III
   B. II, III, IV
   C. III, IV, V
   D. I, II, IV

   Weathering steel is a useful material that contains a small quantity of copper. When exposed to the weather, this material proceeds to rust until it develops a complete protective oxide coating that resists further corrosion. Thus, weathering steel forms its own permanent protection, obviating the need for an additional applied finish or regular maintenance. Among the statements listed, those included in answer B are correct.

2. Exterior plaster is almost always applied in three coats, except in special situations. One of these situations exists when it is applied in two coats over a masonry wall; another such situation is when

   A. the plastered surface is to be completely covered by a veneer.
   B. the finish plastered surface is applied over a base coat that is twice its normal thickness.
   C. the finish plastered surface is applied over an additional continuous surface of metal lath.
   D. the plaster surface is located in a protected area, such as under an overhanging roof.

   Exterior plaster is normally applied over metal lath and consists of a base or scratch coat, a brown coat, and a final finish coat. On monolithic concrete or masonry, the scratch and brown coats may be combined in a single base coat, providing the surface is porous and rough enough to furnish a good bond. The other situation in which two coats of plaster are permitted is when the finish coat is eliminated and replaced by a veneer, such as face brick or stone (correct answer A). No amount of overhang (D), additional lath (C), or increased thickness (B) would be justification to eliminate one of the essential coats of exterior plaster.

   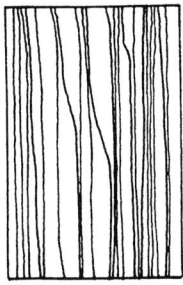

3. If you specify hardwood plywood paneling that has the appearance of the panel shown above, in which of the following ways would the veneer for such plywood be cut?

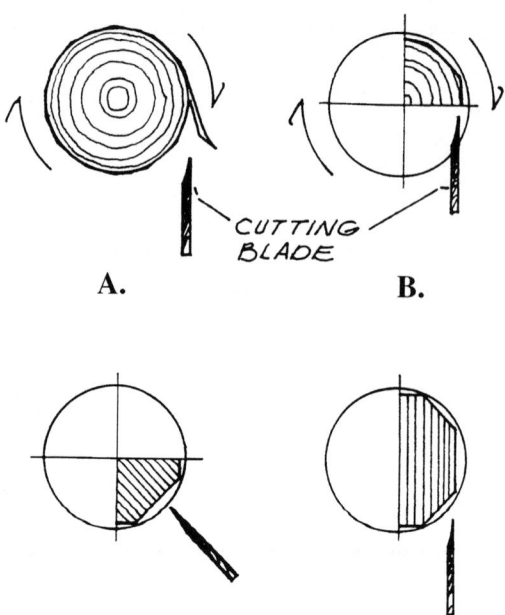

4. During the finishing phase of a building project, ceramic tile is installed using the cement mortar method of application. Before arriving on the job site to oversee the work, the architect should be aware that

   A. the paper-backed, pre-mounted tile should be soaked in water prior to installation.

   B. the loose tile should be soaked in water prior to its installation.

   C. all the ceramic clay tile used on the job should be presoaked prior to installation.

   D. presoaking is an obsolete procedure that could impair the holding power of cement mortar, and thus, should not be permitted.

The way in which veneers are cut determines the ultimate appearance of plywood. Two logs of the same species, when cut differently, will have entirely different characteristics, even though their color and texture may be similar. In the sequence shown, the veneer cutting methods are known as rotary cut (A), rift cut (B), quarter slicing (C), and flat slicing (D). Each of these methods produces the visual characteristics shown below, except for quarter slicing (correct answer C) which is shown on the previous page.

Ceramic tiles are applied by either the cement mortar setting method or the adhesive setting method. The former method is generally preferred because the tile can be applied over the uneven surfaces using the thicker mortar leveling coat. Tile installed with cement mortar is also more durable and generally provides a more water-resistant installation, especially around tubs and showers. Loose tiles intended for mortar beds should be presoaked with water to prevent their drawing water from the cement mortar and weakening the mix (correct answer B). Paper-backed tile should not be soaked, as this may cause individual tiles to separate from the paper backing before the tiles are in place. Incidentally, loose tiles should be damp, not actually wet, at the time they are set.

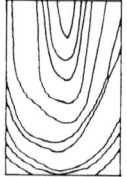 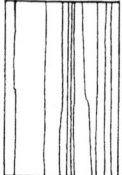 

A.　　　　B.　　　　D.

5. Fire-retardant paints are those that
   I. have a latex base.
   II. have a rubber base.
   III. are oleoresinous.
   IV. are intumescent.
   V. are nonvolatile.

   A. I, V **C.** I, III
   B. II, IV **D.** II, IV, V

Fire retardant paints, often called intumescent paints (IV), are coatings that expand under heat and prevent the flames and heat from spreading to combustible materials. This chemical action, which virtually smothers fire, is caused by the paint's rubber base (II) in combination with other ingredients. These two choices are found in correct answer B. Concerning the other choices, latex-based paint (I) is the most popular kind of interior wall paint; oleoresinous refers to oil paints, which have been largely superseded by alkyd enamels; and nonvolatile is a term applied to ingredients that bind the solid paint particles and provide adhesion to the surface on which they are applied.

HANDICAPPED

1. The goal of barrier-free design is
   A. to permit autonomous functioning.
   B. to permit free use by the nonambulatory.
   C. to eliminate the physical barriers within all buildings.
   D. to provide unobstructed access to all buildings.

Most authorities agree that the goal of barrier-free design is to permit any person with a handicap to participate in normal activities without help—in other words, autonomous functioning (correct answer A). This includes not only the non-ambulatory (B), but also the ambulatory who are handicapped, such as the blind. Choices C and D are also correct, as far as they go, but they are incomplete. Physical barriers must be eliminated outside of, as well as within all buildings (C). And finally, unobstructed access (D) serves little purpose if the handicapped person cannot use the facilities once he arrives at his destination. Candidates should note that in this question, where each choice is partially correct, the right answer is the choice that is most inclusive. In other words, some questions require one to find the answer that is most correct and, therefore, better than all the others.

2. Wristblade levers are generally used for the benefit of handicapped persons on
   A. entrance doors.
   B. lavatory controls.
   C. drinking fountains.
   D. wheelchair controls.

Wristblade type levers are often used on lavatory faucets (correct answer B) for the benefit of handicapped persons. These devices have a winglike projection that enables one to turn water on or off by means of forearm action, rather than by using one's hand. Some handicapped persons, such as quadriplegics or arthritics, find conventional faucets impossible to operate, because the required twisting motion of the hand causes difficulty or pain. Drinking fountains, incidentally, may also employ a lever type control, but not the wristblade variety.

3. Select the statement that is *least* correct.
   A. Barrier-free design is frequently considered to be a civil right.
   B. Barrier-free design is usually beneficial to all building users.
   C. Barrier-free design resolves the life safety problems of the handicapped.
   D. Barrier-free design generally increases a project's cost.

All of the statements concerning barrier-free design are true, with the exception of correct answer C. An unresolved problem exists with handicapped individuals in the event of fire in a multi-storied building. In that case, the able-bodied will exit through enclosed fire-stair towers. The handicapped, however, must rely on elevators, which are risky because they tend to stall at the fire floor; on aid from the able-bodied, which is unpredictable; or on "refuge" compartments at each floor, which are designed to provide protection long enough to allow the fire to be extinguished. All of these alternatives have drawbacks and create a serious problem of life safety that barrier-free design has yet to solve. Concerning choice B, it is true that all building users benefit from certain barrier-free elements that are provided expressly for the handicapped. For example, nearly all able-bodied individuals would appreciate ramps rather than steps, door levers rather than knobs, wider entrances, non-slip floors, and a general absence of hazardous building elements. Incidentally, choice D states that project costs increase with the inclusion of handicapped provisions, which is true, but the increase rarely exceeds 2 percent, on the average, of the total project budget.

4. A barrier-free environment is one in which there is
   A. an absence of steps, stairways, and sloped walks exceeding 10 percent.
   B. an absence of any obstruction to mobility or function.
   C. consideration for those confined to a wheelchair.
   D. consideration for all those having a disabling condition.

   To one degree or another, each of the choices defining barrier-free is correct; however, only the correct answer B is complete. A barrier-free environment may, of course, have steps or slopes (A) for the use of able-bodied individuals, but such an environment must also have ramps, adequately-sized doorways, toilet stalls, parking spaces, and a great many other elements required by the handicapped. Barrier-free also applies to many individuals other than those confined to a wheelchair (C). Handicapped people include those who are frail because of age, on crutches because of a broken leg, or even excessively fat, very short, or merely pregnant. Concerning choice D, there are some disabling conditions that are not classified as handicaps. For example, people who are deaf or epileptic may have a medically-defined disabling condition, but it is not the kind of condition that will prevent them from overcoming most common architectural barriers.

5. In a public building it is necessary to provide an appropriate number of public telephones that are accessible to physically handicapped persons. In this regard, which of the following features should be incorporated?
   I. 36 inch minimum square booths
   II. 48 inch maximum coin slot height
   III. 48 inch maximum dial height
   IV. Provisions for hearing disabilities.
   V. Visual identification sign
   A. I, II, III   C. I, II, V
   B. III, IV, V   D. II, III, IV, V

   All of the features listed should be incorporated when providing barrier-free phones, with the exception of I. Telephones should not be placed in phone booths; they should be wall-mounted instruments that provide easy access to those in wheelchairs. Furthermore, if a phone compartment were provided, it would have to be at least four feet square and have a non-obstructing door. All the other features are standard practice, as expressed in correct answer D.

6. Barrier-free parking spaces are always wider than regular parking spaces because
   A. handicapped people invariably drive full size, rather than compact cars.
   B. handicapped people often have impaired coordination in driving, which requires an allowance for greater parking tolerances.
   C. additional space must be allowed for those who assist the handicapped in entering or exiting their cars.
   D. additional space must be allowed for the mobility of those in wheelchairs or on crutches.

Spaces for handicapped persons are generally 12 to 13 feet wide, which is about half again as wide as regular parking spaces. The reason for this additional width is accurately stated in correct answer D; individuals using wheelchairs, braces, or crutches simply require more than normal space to maneuver between parked cars. Incidentally, there is absolutely no evidence to support statements A or B. In fact, those with impaired coordination are not permitted to drive cars. Finally, in choice C there is the assumption that handicapped people who drive are unable to get in or out of their cars without help. This, too, is a misconception.

7. A knurled door knob would be of greatest benefit to someone who is

   A. blind.
   B. arthritic.
   C. confined to a wheelchair.
   D. using crutches.

Knurled door knobs are those which have a series of small ridges carved around their edges. They are used principally to identify hazardous areas to those who are visually handicapped (correct answer A) and, thus, cannot be warned by other means. A blind person touching the knurled surface is alerted to the fact that beyond the door may lie a boiler room, fire escape, loading platform, or other area that is a hazard to one who is unable to see. For this same purpose, square knobs are sometimes used, as well as knurled levers, etc.

8. The minimum turning radius of a typical wheelchair is approximately

   A. 18 inches   **C.** 32 inches
   B. 24 inches   **D.** 42 inches

Because a wheelchair so frequently characterizes the problems of the handicapped, candidates should know its basic size and operating dimensions as thoroughly as one knows the pertinent facts about cars. A wheelchair is approximately 25 inches wide, wheel to wheel; about 36 inches high, from the floor to the push handles; and it has a fixed turning radius, when pivoting on a spot, of between 32 and 36 inches. Answer C, therefore, is correct.

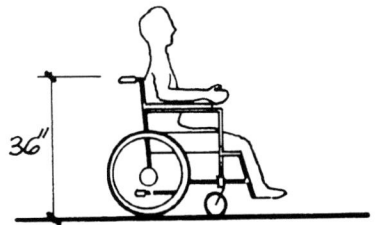

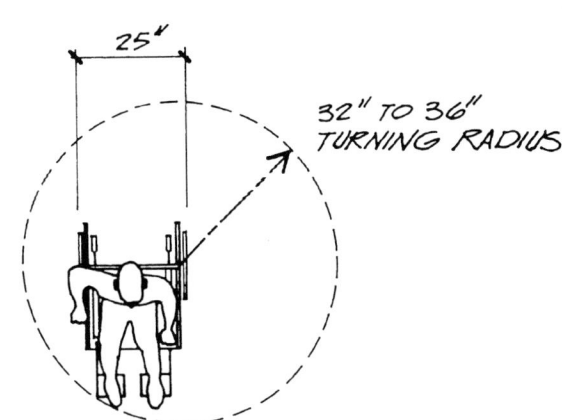

9. Rank the following types of walkway surfaces, from least to most desirable, to accommodate the physically handicapped.
   I. Flagstone
   II. Asphalt
   III. Tanbark
   IV. Brick in concrete
   A. IV, II, III, I
   B. II, IV, I, III
   C. I, II, IV, III
   D. III, I, IV, II

The surface of walkways should be stable and firm, relatively smooth in texture, and have a nonslip quality in order to allow barrier-free movement for the handicapped. The architect should, of course, consider other characteristics in the selection of materials, such as maintenance and installation costs, ice and snow removal, and erosion. This question asks us to evaluate the surface characteristics that provide safety and ease of movement for wheelchairs and people with mobility handicaps. Because of its irregular texture and relative softness, tanbark is least desirable in such applications. Although flagstone provides a hard surface, the irregularity of the shapes and resultant wide joints make walking extremely difficult for people with mobility handicaps. Furthermore, the irregular surfaces of such a natural material make movement difficult for wheelchairs. Asphalt and brick in concrete are about equally good hard surfaces for walkways, although asphalt has no joints and, therefore provides the best possible solution where handicapped mobility is a primary consideration. The correct answer is D.

HVAC

1. In the compression refrigeration cycle, the pipe connecting the evaporator to the compressor contains
   A. a medium temperature, high pressure liquid.
   B. a low pressure, low temperature gas.
   C. a high pressure, high temperature gas.
   D. a high pressure, low temperature liquid.

In the compression refrigeration cycle, the refrigerant leaving the compressor for the condenser is a high pressure, high temperature gas. Between the condenser and the evaporator it is a high pressure, medium temperature liquid, and after boiling in the evaporator at a low pressure, it leaves the evaporator for the compressor as a low pressure, low temperature gas. Therefore B is the correct choice.

2. A wall has a U value of 0.20. In order to reduce heat transmission through the wall, 1 1/2 inches of insulation having a K value of 0.30 is added. What are the R and U values of the wall after adding the insulation?
   A. R = 1.54, U = 0.65
   B. R = 2.0, U = 0.50
   C. R = 10.0, U = 0.10
   D. R = 8.33, U = 0.12

In order to answer this question, a candidate must understand the terminology: U, K, and R. The U factor is the overall coefficient of heat transfer of a building section (floor, roof, or wall), and is equal to the number of BTU per hour that will pass through one square foot of the section when the temperature difference is 1°F. Thus, the lower the U value, the less heat transferred, and therefore, the better its insulation characteristics. The K factor is the coefficient of heat transfer of a material exactly one inch thick, and is equal to the number of BTU per hour that will pass through one square foot of the material per inch of thickness when the temperature difference is 1°F. The lower the K value, the less heat transferred, and thus the better the insulation properties of the material. R is the resistance to heat transfer of a material or a building section. When referring to a material, R is the reciprocal of K (R = 1 ÷ K), and when referring to an overall building section, R is the reciprocal of U (R = 1 ÷ U). The greater the R value, the less heat transferred, and therefore the better the insulating properties of the material or the section. The original U value of the wall is 0.20. Since R = 1 ÷ U, R = 1/0.20 = 5.0. Now we add 1 1/2 inches of insulation having a K value of 0.30. R for the insulation is equal to 1/K 1/0.30 = 3.33 per inch of thickness. Since the insulation is 1 1/2 inches thick, its total resistance R = 1.5 × 3.33 = 5.0. The total R for the wall plus the 1 1/2 inches of added insulation is 5.0 + 5.0 = 10.0. The combined U factor is 1 ÷ total R = 1/10.0 = 0.10. C is the correct answer.

3. The return duct of a centralized air conditioning system
   A. returns the air from the air conditioned rooms back to the central supply fan.
   B. aids the supply fan in providing air to the distant corners of the building.
   C. exhausts air from the building.
   D. provides outside air to the building.

The return duct provides the means of recirculating (returning) the room air back to the supply fan, to be cooled or heated as required. Therefore A is correct.

4. The dew point is reached at
   I. 100 grains of moisture per pound of dry air and 100 degrees dry bulb.
   II. 60° DB and 60° WB.
   III. 100 grains of moisture per pound of dry air and 100 degrees wet bulb.
   IV. 100 percent RH.

   A. I and II   **C.** III and IV
   B. II and III   **D.** II and IV

The dew point is the temperature at which air can hold no more moisture in solution; i.e., the air is saturated. This would correspond with identical dry bulb and wet bulb temperatures, since no moisture could evaporate from the wet bulb thermometer to depress the wet bulb temperature. The dew point temperature would also correspond with 100 percent relative humidity, as the definition of relative humidity is the amount of moisture in the air relative to the amount it could hold if saturated. Therefore D is correct.

5. The best method to reduce solar heat gain through windows is with the use of
   A. venetian blinds.
   B. exterior shutters.
   C. draperies.
   D. double-glazing.

Since the object in reducing heat gain is to prevent solar energy from entering the building, it is always most effective to stop the radiation on the outside, before it reaches the glass. Therefore choice B is correct, as exterior shutters will shade the glass. Double-glazing reduces conduction gains and losses, but does very little to reduce solar radiation unless the glazing is reflective. Finally, venetian blinds and draperies can do little more than trap the heat against the glass once it enters the building. At that point, however, solar heat gain has penetrated the space.

6. In a ten story building, the water chiller, hot water boiler, chilled and hot water pumps are in the basement. The HVAC fans and coils are on the fifth floor. The cooling tower and condenser water pump are on the roof. Which of the following mechanical services would be found in a shaft through the seventh floor?
   I. Supply and return air ducts
   II. Chilled water piping
   III. Hot water piping
   IV. Condenser water piping

   A. I and IV   **C.** I, II, III, IV
   B. I and II   **D.** II, III, IV

Supply and return ducts would run between the first floor ceiling and the roof. Chilled water and hot water piping would run between the equipment in the basement and the coils on the fifth floor. Condenser water piping would run between the cooling tower on the roof and the chiller in the basement. Therefore supply and return ducts and condenser water piping would pass through the seventh floor; A is therefore correct.

7. Heat always flows
   A. upward.
   B. horizontally.
   C. hot to cold.
   D. cold to hot.

The basic law of thermodynamics is that heat always flows from hot to cold. A heated gas, such as air, flows upward by convection; however heat will conduct and radiate in any direction, but always from hot to cold. Therefore C is the correct answer.

8. Degree-days are used to determine
   A. the heating requirements of a building.
   B. the design temperature of a space.
   C. the average daily temperature at a site.
   D. the solar heat gain at a site.

A degree-day is the amount by which the average outdoor temperature at a particular location is below 65°F for one day. The number of degree-days per month or per year is the sum of the degree-days for each day in that period. Thus, a cold region has a large number of degree-days per year, while a warmer region has fewer degree-days per year. For example, Alaska averages 10,000 to 20,000 degree-days per year, northern Minnesota around 10,000, and southern Florida less than 500. The heating requirement of a building for any period of time is directly proportional to the number of degree-days for that period (correct answer A).

9. Which of the following best describes a VAV system?
   I. Conserves fan energy during light loads.
   II. Modulates the supply air temperature very closely.
   III. Requires no return air duct.
   IV. Is easily adaptable to changing zones or adding zones in the future.
   A. I, II, III, IV
   B. II, III, IV
   C. I and II
   D. I and IV

Since a VAV (variable air volume) system varies the volume, rather than the temperature, of supply air, at light loads the volume of air supplied will be reduced, thus reducing fan work and conserving fan energy. Therefore I is correct and II is not correct. Since VAV systems are supplied from centralized fans, return air ducting is necessary (III is incorrect). As for future adaptability, adding zones to a VAV system merely requires tapping into the main supply duct with an additional VAV unit (IV is correct). The correct answer is D.

10. Heat transfers across a vacuum by means of
    A. radiation.
    B. conduction.
    C. convection.
    D. none of the above.

Conduction is the method of transferring heat through a substance by passing from one particle to another. Convection is heat transfer by moving currents of air or any other gas or liquid. In a vacuum, there is no medium available to conduct heat or to create convection. We know, however, that heat transfers across a vacuum, because we receive heat from the sun across the vacuum of space. This method of heat transfer is radiation (correct answer A), which is the method of heat transfer in which heat travels rapidly in straight lines without heating the intervening space.

11. Which of the following best describes a ton of refrigeration?
    I. 10,000 BTU per hour.
    II. 12,000 BTU per hour.
    III. The amount of heat necessary to melt a ton of ice over a 24-hour period.
    IV. The amount of heat extracted by a refrigeration unit weighing 2,000 pounds.
    A. I and III
    B. I and IV
    C. II and III
    D. II and IV

A ton of refrigeration is defined as the amount of heat necessary to melt (change state from a solid to a liquid) a ton of ice over a 24-hour period. The latent heat of fusion of water, which is the heat required to change one pound of ice to one pound of water, is 144 BTU. To find the BTU per hour equal to one ton of refrigeration, 144 BTU/lb × 2,000 lbs ÷ 24 hours = 12,000 BTU/hr. Therefore C is the correct answer.

12. Which of the following best describes a two-pipe system?
    A. Hot water flows in one pipe and chilled water flows in the other pipe.
    B. The north side of a building can be heated and the south side can be cooled simultaneously.
    C. Requires one pump only.
    D. Requires a hot water and chilled water pump.

A two-pipe system handles either hot water supply and return or chilled water supply and return. The single pump pumps the water either through a hot water boiler or a water chiller, depending upon which is required. Since only hot or chilled water is flowing, the building cannot be heated and cooled simultaneously. Therefore C is the only correct answer.

LEGAL & ZONING

1. Many construction contracts contain a liquidated damages clause, which provides that if the contractor fails to complete the work by a certain date, a stipulated amount will be deducted from the contract price for each day of delay. Select the correct statement about liquidated damages.

   A. A liquidated damages provision is usually enforceable even if the amounts stipulated are greatly in excess of the actual damages to the owner caused by the delay.

   B. If the owner allows the contractor to work after the expiration of the contract completion date, he automatically waives his right to liquidated damages.

   C. If the specifications require the use of a material only available from one source, any resulting delay is the responsibility of the owner and the liquidated damages provision is therefore unenforceable.

   D. Even though a contract contains a liquidated damages provision, it may also contain a provision for extending the time for completion.

This may sound like a question from the bar exam, but some knowledge of construction contracts and law is also required of architects. Let's analyze each choice. In choice A, if the amount stipulated in a liquidated damages provision is excessively disproportionate to the actual damages, it will be considered a penalty rather than liquidated damages, and will not be enforced by the courts. A is therefore false. B is also a false statement. The liquidated damages provision assumes that the contractor will continue the work after the expiration of the contract completion date, if he fails to complete the work by that date. The difficulty of obtaining the specified material in choice C does not relieve the contractor of his obligations under the contract. Under those circumstances, the liquidated damages provision remains enforceable, making C a false statement. D is true and the correct answer to this question. At the discretion of the owner, the time for completion may be extended for delays over which the contractor has absolutely no control.

2. During the course of construction, temporary formwork erected by the contractor collapses, injuring a workman. The injured workman files suit against the architect, claiming that the architect was negligent in supervising the work, since he did not notice or call attention to the faulty forms. Which of the following is the most correct statement concerning this situation?

   A. The suit will be dismissed, since no contractual arrangement exists between the architect and the workman.

   B. Whether or not the suit has merit, the architect's professional liability insurance company will defend the architect, thus avoiding any out-of-pocket expense to him.

   C. The claim of the injured worker will be paid by the contractor's workers' compensation insurance, rather than the architect's professional liability insurance.

   D. Although the General Conditions place responsibility for construction methods on the contractor and not the architect, the architect is not automatically absolved of negligence and he must still defend himself against the suit.

This is a question testing your knowledge of architects' liability and professional liability insurance. Choice A is false. Until the early part of this century, actions for negligence against architects were usually brought only by persons with whom they had contracted. In more recent years, however, the courts have extended the liability of architects to include injured third parties. The first part of statement B is correct—his insurance company will defend him. However, since almost all professional liability insurance has a large deductible amount, including defense costs, the architect must pay part or all of the costs of his defense. B must therefore be considered false. The injured workman is covered by workers' compensation insurance, which provides prompt compensation for his job-related injury. However, this does not prevent him from seeking additional recovery from the architect or other parties. C is therefore false. D is a true statement and therefore the answer to this question; the architect must defend himself against the suit, with the court making the final determination concerning his negligence.

3. In general, why does a governmental agency use the right of eminent domain?
   I. To acquire land from an owner who refuses to sell at any price.
   II. To acquire land from an owner who refuses to accept the purchase price offered by the agency.
   III. To establish a scale of values, which is then used to determine fair compensation for all owners whose property is acquired.
   IV. To avoid the normal procedures for buying land, which are costly and time-consuming.
   V. To avoid a court determination of the value of acquired land.

   A. I, II, IV
   B. I and II
   C. I, III, V
   D. II, IV, V

Eminent domain is the right of the government to take private property for public use, with the owner receiving fair compensation. Ordinarily, a governmental agency will first attempt to buy the land in the normal manner, negotiating the price with the owner, rather than exercise its right of eminent domain (incorrect statements III and IV). If agreement with the owner cannot be reached, eminent domain may be used as a last resort (I and II). In that case, the value of the property may have to be determined by the court (V). Since statements I and II are correct, B is the answer to this question.

4. An architect designed a racetrack pavilion in the early 1960s. His design provided for plate glass windows that would allow patrons to watch the races and yet be protected from the weather. Twelve years later, a child outside the pavilion area threw a rock that hit one of the windows and shattered it. A patron sitting nearby was showered with glass fragments, causing very minor cuts. Two years after this incident, a suit was filed against the architect and others, alleging that the architect was negligent in not specifying safety glass and in placing seating too close to the windows. Damages in the amount of $350,000 were demanded. What is the most likely outcome of this suit?

   A. The architect would lose the suit and be liable for $350,000 in damages.
   B. The architect would win the suit, and the plaintiff's attorney would be held liable to the architect for his costs of defense.
   C. The architect would win the suit, but would have to bear the burden of his defense costs.
   D. The architect would lose the suit, but the court would reduce the award to less than $350,000.

Like many questions on the exam, and in real life, this question has no absolutely right or wrong answer. What we are looking for is the most correct answer, the most likely result of the lawsuit. Let's review the salient facts: 1) The incident occurred 12 years after construction, and the suit was filed two years later. 2) The glass specified was apparently strong enough to withstand wind forces, but not the impact of a rock thrown at it. 3) The injuries sustained were minor. Now let's eliminate the least likely answers. It is very unlikely that any court would award $350,000 to the injured party, given the facts as presented, and thus, we rule out answer A. B can also be eliminated, since it is unlikely that the plaintiff's attorney would be held liable for the architect's defense costs. The reason? Public policy favors free access to the judicial process, and consequently, attorneys traditionally enjoy special protection in the courts. It comes down to C or D—would the architect win or lose the suit? In the actual court case on which this question is based, the court decided in favor of the architect (correct answer C). The great length of time between construction and filing of the suit (14 years), and the adequacy of the glass under normal service conditions were factors that helped absolve the architect of any negligence. However, he still had to bear the costs of his defense.

5. Match each zoning term with the appropriate description.

    I. Spot zoning
    II. Variance
    III. Conditional use
    IV. Nonconforming use

    1. Special permission to deviate from normal zoning requirements, for a special purpose such as a school or hospital.
    2. The designation of a parcel of land for a use classification different from that of the surrounding area, to favor a particular owner.
    3. Special permission granted to an owner to deviate from the zoning requirements normally applicable to his property.
    4. A use not complying with the zoning ordinance, but permitted because it predates that ordinance.

    A. I-2, II-3, III-1, IV-4
    B. I-1, II-3, III-4, IV-2
    C. I-4, II-2, III-1, IV-3
    D. I-3, II-1, III-2, IV-4

A zoning ordinance is a law enacted by a municipality for the general welfare, which establishes zones or districts, within which the location, height, and use of buildings are regulated. Each of the four terms above represents a deviation from normal zoning regulations. If a building already exists when a zoning ordinance is enacted, its lawful use may usually be continued even though it does not conform with the zoning regulations. This is known as a nonconforming use (IV-4). Where strict application of zoning regulations would result in exceptional hardship, special permission to deviate from the regulations may be granted. Such permission is called a variance (II-3). A conditional use is similar to a variance, except that it permits a special use, such as a school or hospital, which is for the public welfare and convenience (III-1). Sometimes, small spot zones are established, which may be inconsistent with the general intent of the zoning regulations and which favor a particular property owner (I-2). The terms are correctly matched in answer A.

6. Select the correct statement.
   A. Cluster zoning permits a higher overall density than conventional zoning.
   B. A PUD requires a zoning variance.
   C. Cluster zoning results in permanent community open space.
   D. A PUD is limited to residential development only.

Let's examine the two terms that appear in this question: cluster zoning and PUD. In cluster zoning, a developer is permitted to reduce the minimum lot size below that required for conventional zoning, so long as the total number of dwellings in the subdivision remains the same (incorrect statement A), and if the land gained thereby is preserved as permanent community open space (correct statement C). A PUD (Planned Unit Development) is similar to a cluster development, but is larger in scale and may include commercial and industrial development, as well as housing (incorrect statement D). Since a PUD utilizes the cluster zoning concept, it generally does not require a zoning variance (incorrect statement B). C is therefore the answer.

7. In cases involving construction, the period of time within which legal action must be brought for alleged damages or injury is established by a
   A. statute of frauds.
   B. statute of limitations.
   C. lien.
   D. time is of the essence clause.

Common to most systems of justice is the concept that there should be a time limit to the right to bring suit against an individual. In this country, a law which establishes such a time limit is called a statute of limitations (correct answer B). Almost all civil law actions are covered by some form of statute of limitations, and these vary from state to state and with the type of legal action. Normally the period commences with the discovery of the act resulting in the damage or injury; but when does the statute start to run on a construction project? When the construction is completed? Or when the damage is discovered? Again, the statutes vary widely among the states. Incidentally, both design professionals and contractors are subject to legal action if injury or damage results from faulty workmanship or defective design. Looking at the other choices, a statute of frauds (A) is an early English statute, recognized by most states, which provides that real estate contracts must be in writing to be enforceable. A lien (C) is a claim on property which encumbers it until the obligation is satisfied. And finally, a time is of the essence clause (D) is a clause in a contract which makes the date of completion of the contract terms an essential element of the contract.

8. The Ajax Electrical Supply Co. furnishes equipment to Watts and Voltz, a subcontractor, to be used in two major construction projects as follows: $50,000 in equipment for a public school project and $40,000 in equipment for a private commercial project. Six months after the materials have been furnished, Watts and Voltz goes bankrupt. Ajax has not been paid for any of its materials and is listed as a creditor of Watts and Voltz. Which of the following statements is most correct?

   A. Ajax has no recourse against either owner.
   B. Ajax only has recourse against the private owner under mechanic's lien laws.
   C. Ajax has recourse against both owners under mechanic's lien laws.
   D. Ajax only has recourse against the public school district under mechanic's lien laws.

A mechanic's lien is a statutory lien on real property for payment of work performed or materials furnished for the improvement of real estate. Although mechanic's lien statutes vary from state to state, their general characteristics are the same. The right to a mechanic's lien is available to anyone who has furnished labor or materials for a construction project for which he has not been paid, provided he has complied with all the requirements of the governing mechanic's lien law. Among those protected by mechanic's lien laws are architects, contractors, laborers, material suppliers, and subcontractors. However, in general, a mechanic's lien does not attach to and cannot be enforced against public property owned by the federal government or a state, county, city, or other public district. The correct answer is therefore B.

MASONRY

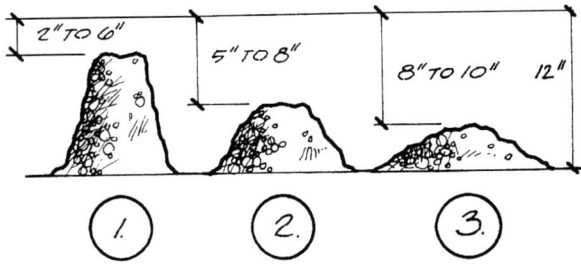

1. Shown above are the results of three different slump tests. In the same order as shown, one should know that the materials indicated are

    A. concrete, grout, and mortar
    B. concrete, mortar, and grout
    C. exterior plaster, grout, and concrete
    D. mortar, grout, and concrete

Concrete and grout are both composed of varying proportions of cement, water, sand, and gravel. Mortar is similar in composition, but instead of gravel, lime is used. Therefore, all three materials are much alike, except for their plasticity or fluidity in the initial stage. Since the slump test is used to measure the consistency of a mix, one should know that concrete, with a tightly controlled water-cement ratio, is relatively stiff; mortar, which must be handled by a trowel, is less stiff; and grout, which is poured between wythes in a brick wall or into the cells of hollow concrete masonry units, is actually quite fluid. The correct answer, therefore, is B.

2. In a large project calling for reinforced hollow masonry units, the owner has decided to substitute common brick, because of its color, texture, and scale. Under these circumstances, the architect is obliged to advise the owner that this change will undoubtedly

    A. cause a problem in accommodating the steel reinforcing.
    B. make very little difference in the overall construction budget.
    C. make it difficult to maintain the original construction schedule.
    D. result in a structure that is incapable of resisting the calculated design loads.

In masonry construction, there are obvious differences between using block and brick; and in this question, candidates must determine what those differences are. To begin with, there should be no problem accommodating the reinforcing bars (A), since a reinforced brick wall is composed of two wythes between which the steel bars can be placed at almost any spacing. Statement B is also false, because brick is invariably more costly than concrete block. More units are involved requiring more handling, and two wythes must be constructed requiring a greater amount of time and thus, money. These two facts are also the reasons that statement C is the correct answer. The handling of many more units simply takes more construction time. Finally, statement D is false, since reinforced brick walls should be able—in most cases—to resist loads similar to those resisted by reinforced hollow masonry units.

3. Skintled brickwork would be best used in
   A. areas with freezing climates.
   B. areas with abundant rainfall.
   C. commercial structures.
   D. garden walls.

   It is an uncommon term perhaps, but "skintled" is not unlike other such relatively obscure terms that have been tested on past exams. The word refers to a pattern of brickwork in which the faces of some bricks are set out of line with the face of the wall. The mortar that protrudes is often allowed to remain, forming a squeezed joint. The result is an irregularly-faced wall that has a rugged appearance, and also one that frequently leaks and collects dirt. Therefore, if one were to use skintled brickwork at all, one would be wise to confine its use to a garden wall (correct answer D).

4. Select the correct statement, concerning masonry joints, from among those that follow.
   A. The best type of joint to resist water penetration is either a weather-struck joint or a trowel-struck joint.
   B. The best type of joint to use on a masonry wall that is to receive a plastered finish is a flush joint.
   C. The type of joint that can be easily formed with a trowel is a vee-shaped joint.
   D. The types of joints used on interior masonry walls are essentially the same as those used on walls exposed to the weather.

   All of the statements concerning masonry joints are false, with the exception of correct answer B. Statement A is actually only half wrong, since a weather-struck joint sheds water, but a trowel-struck joint is an invitation to a leaky wall. The vee-shaped joint of statement C is formed not with a trowel, but with a special jointing tool that compresses the mortar and forms a weathertight seal. Finally, choice D is simply a false statement; any wall that is not exposed to the weather can employ any kind of joint one desires. Since it need not shed water, the choice becomes one of aesthetic preference.

5. A masonry wall employing Norman-sized modular bricks is proposed, employing joints that are approximately 3/8 inch thick. What will be the height of three courses in this wall?
   A. 8 inches
   B. 9 inches
   C. 12 inches
   D. 16 inches

   Although efforts have been made to standardize brick sizes, the industry still maintains literally dozens of different sized masonry units that create confusion among designers and builders. Modular bricks were developed to regulate coursing dimensions. There exist about five modular brick heights, in a variety of widths and lengths, which, together with a mortar joint thickness of between 1/4 and 1/2 inch, result in even coursing dimensions. A course of Norman brickwork, for example, is 2-2/3 inches high, 4 inches wide, and 12 inches long. When laid in a wall, three courses of Norman modular brick therefore are exactly 8 inches high (correct answer A).

METALS

1. Select the properties or characteristics applicable to light-gauge punched steel studs.
   I. Cannot be nailed directly
   II. Easy passage of pipes
   III. Unaffected by rot, warpage, shrinkage, or termites
   IV. Incombustible
   V. One-hour fire resistance regardless of finish materials used

   A. II, III, IV
   B. I, III, IV, V
   C. II, V
   D. All of the above

   Light-gauge steel shapes for the framing of buildings, including studs and joists, possess several distinct advantages. They do not swell, shrink, or warp, and they are unaffected by termites or rot (III). They can be used where incombustible construction is required (IV). When punched, steel studs permit the easy passage of pipes and conduits (II). Steel studs are often furnished with a nailing groove, allowing finish materials to be nailed directly (I is incorrect). The fire-resistive period of a steel stud wall is determined largely by the finish materials used (V is incorrect). The correct characteristics are II, III, and IV (correct answer A).

2. Steel is often finished for protective or decorative purposes with a variety of metallic coatings. In this connection, match each term with the appropriate description.
   I. Hot dipping
   II. Electroplating
   III. Metallizing
   IV. Bonderizing
   V. Sherardizing

   1. Deposition of a metallic coating on steel by means of electrolysis.
   2. Applying a corrosion-resistant phosphate coating to steel, which serves as a base for paint.
   3. Placing steel in an enclosure in which it is surrounded with zinc dust and then heated.
   4. Immersion of steel in a bath of molten metal.
   5. Spraying molten metal on the surface to be coated.

   A. I-3, II-1, III-4, IV-2, V-5
   B. I-4, II-3, III-1, IV-5, V-2
   C. I-2, II-1, III-3, IV-4, V-5
   D. I-4, II-1, III-5, IV-2, V-3

   Metallic coatings provide protection against corrosion either by being a sacrificial metal—that is, one that is purposely permitted to corrode, as is the case with zinc coatings; or by being inert and corrosion-resistant, as with chromium. The metals used for protection include zinc, tin, terne, aluminum, cadmium, chromium, and nickel. Some of the coating processes used are those listed above, which are described as follows: Hot dipping refers to the immersion of steel in a molten bath of the coating metal (I-4). This process is used to produce heavier coatings, especially galvanized sheets, where the protective metal is zinc. Electroplating employs an electric current and an electrolytic solution to deposit a metallic coating on steel (II-1). Tin, zinc, and cadmium are commonly applied in this manner. Lighter coatings for minimum corrosion resistance are possible with electroplating. Metallizing is the process of spraying molten metal on the surface to be coated (III-5). It is extensively used for applying zinc and aluminum coatings, and is the principal method used for field application. Bonderizing produces an excellent paint base by dipping in a hot phosphate (IV-2). Finally, the sherardizing process consists of placing steel into an enclosure in which it is surrounded with metallic zinc dust and then heated (V-3). This process, which is limited to relatively small objects, produces a clear, thin zinc coating. The terms and descriptions are correctly matched in answer D.

3. Steel floor decking is widely used because of its great strength, light weight, and speed of construction. Which of the following statements concerning steel decking is true?

   A. Because of the smooth finish of steel decking, composite action of the decking and the concrete floor slab is not permitted.

   B. The most typical method of attaching steel decking to the supporting framework is welding.

   C. When a concrete floor fill is placed over steel decking, no additional fireproofing of the steel decking or frame is required.

   D. Steel decking is often used for electrical raceways, because it provides easy access from below.

Certain types of steel decking are specifically designed to provide composite action with the concrete floor slab (false answer A). What makes this action possible are special deformations formed into the decking that bond it to the concrete and transfer horizontal shear between the decking and the concrete. B is the true statement we are looking for and therefore the answer to this question. Although other means of attachment are sometimes used, welding remains the typical method of attachment. Even when a concrete floor fill is placed over steel decking, the steel frame and usually the underside of the deck must be fireproofed to obtain the required fire-resistive rating. C is therefore false. Finally, D is also a false statement. The easy access to the electrical raceways provided by steel decking is from above, not below.

4. Which of the following statements about aluminum is *incorrect*?

   A. One of the principal advantages of aluminum is its resistance to oxidation.

   B. Aluminums used in building construction are usually alloys.

   C. Some of the important properties of aluminum include light weight, high thermal and electrical conductivity, and easy workability.

   D. Aluminum in contact with other metals is subject to galvanic corrosion when moisture is present.

Aluminum combines readily with oxygen (oxidation) to form a thin film of aluminum oxide, which resists corrosion. A is therefore false and the answer to this question. Virtually all aluminums used in building construction are alloys (B) of pure aluminum with small amounts of iron, silicon, and other elements. Statement C correctly lists some of the important properties of aluminum. Finally, statement D is also correct; aluminum in contact with other metals when moisture is present can corrode as a result of galvanic action. Proper measures to isolate the aluminum from these metals should therefore be taken.

5. Select the correct statement from those that follow.

   A. Steel nails can be used to fasten stainless steel without any concern about galvanic action.

   B. Conduits and pipes of aluminum are often used in reinforced concrete construction, because the concrete forms an effective barrier preventing galvanic action between aluminum and steel.

   C. All ferrous metals contain some silicon, and small variations in silicon content have an important influence on their properties.

   D. Steel can be formed by a variety of methods, but not extruded.

Steel nails should not be used to fasten stainless steel (A), because the stainless steel will cause the steel nails to deteriorate through galvanic action. Unless effectively coated or covered, the use of aluminum in structural concrete is prohibited (B), because aluminum not only reacts electrolytically with steel, but also reacts with the concrete. Although ferrous metals may contain silicon, their properties are largely influenced by their carbon, not silicon, content (C). Finally, steels can be wrought, cast, rolled, forged, and welded, but not extruded (D). Extrusion refers to forming metal by forcing it through an opening having the desired shape. D is the correct answer to this question.

MOISTURE PROTECTION

1. Cant strips are often installed where a roof surface meets an adjoining wall. The principal reason for using this device is so that
   A. roofing sheets may be turned up the wall at a more moderate angle than 90 degrees.
   B. sheet metal flashing will not have to be bent at an angle greater than 90 degrees.
   C. roof water will not puddle at the critical joint between the roof and wall planes.
   D. allowance for movement due to expansion and contraction is provided, with no loss of weathertightness.

Cant strips are commonly used in connection with the flashing of built-up roofs. At the critical roof-wall joint, L-shaped sheet metal flashing is used, or an alternate method is employed in which the roofing sheets are run up and over a 45 degree angle cant strip and sealed by flashing along the wall. The cant strip enables the roofing sheets to be bent at a more moderate angle (correct answer A), which lessens the possibility of their failure. Choices B and C do not apply to this situation, and statement D actually refers to an expansion joint, which is quite different from a cant strip.

2. In the design of metal gutters used to drain a sloping roof, one must be concerned with
   I. annual precipitation
   II. roof pitch
   III. roof area
   IV. cross sectional gutter area
   A. I and II
   B. II, III, IV
   C. I, III, IV
   D. I, II, III, IV

Candidates may consider this to be a deceptive question, because all the factors listed are important, with the exception of I. The amount of rain that falls during a year has no real influence on the design of gutters. It is the intensity and frequency of precipitation that determines the capacity of gutters; in other words, how much rain will fall in a short period of time, and how often will this occur? Roof area and pitch will determine the amount of roof water runoff, and the gutter area will be a function of this runoff quantity. The right combination of factors is found in correct answer B.

3. Mastic sealers are used to seal joints, gaps, or cracks that cannot otherwise be made completely tight. Therefore, they must possess all of the following qualities, *except*
   A. remaining sufficiently flexible to permit movement between the two surfaces.
   B. remaining free from cracks and blisters or exuding during periods of high temperature.
   C. developing rapid bond with high strength while remaining durable over long periods of time.
   D. being suitable for application by hand, trowel, or pressure gun.

Mastic sealers have all of the qualities listed, with the exception of correct answer C. Rapid bond and high strength are properties that more properly belong to adhesives, which are distinctly different from mastics. It is true that mastics must also adhere to surrounding materials, but their fundamental difference is that mastics are intended to seal, that is, keep out water, whereas the purpose of adhesives is to join two materials.

4. A small building has been designed with roof planes that slope at a pitch of 4:12. It was originally intended that wood shingles would be used as a roof covering; however, in an effort to reduce construction costs, the client now insists on using asphalt shingles. It should be explained to the client that this seemingly simple modification will necessitate a change in the

   A. roof slope.
   B. roof sheathing.
   C. type of roofing attachment used.
   D. type of roof flashing used.

Wood shingles and asphalt shingles are actually quite similar in several respects. Essentially they are both applied in an overlapping fashion on any roof surface that has sufficient slope to insure proper drainage. Specifically, a slope of 4:12 is considered a safe minimum slope for both shingle types (A). Both types of shingles also require similar attachments and flashing (C and D), such as hot-dipped galvanized steel or aluminum roofing nails and galvanized, aluminum, or copper flashing. The necessary change, therefore, would be in the roof sheathing (correct answer B). The use of wood shingles requires spaced sheathing strips, to permit air circulation behind the units, whereas asphalt shingles must be applied over solidly sheathed roofs.

5. The primary purpose of installing a moisture vapor barrier is to

   A. lower the relative humidity within a structure.
   B. make a structure moisture proof.
   C. prevent condensation within a structure.
   D. prevent heat loss from a structure.

Moisture in the form of vapor is present in all air, and the quantity of this vapor increases with the temperature of the air. When the air reaches its saturation point, its temperature is referred to as the dew point. At the dew point, vapor condenses and appears as moisture. Moisture vapor flows from areas of high temperature to areas of low temperature (high to low pressure), and because of this flow, moisture vapor is capable of passing through porous construction. When passing through a structure, the vapor comes in contact with colder elements, it falls to its dew point, and it condenses as water within the structure, often within wall or ceiling insulation. The result is decay or damage to building elements. Vapor barriers are intended to prevent vapor flow to areas where such damage from condensed moisture can occur (correct answer C). The barrier itself, usually formed from plastic films or metal foils, must be at a temperature above the dew point, therefore, it is always installed on the warm or room side of the construction. Incidentally, answer D is not completely untrue; however, heat loss prevention is a secondary purpose of a vapor barrier.

6. Flat built-up roofs frequently have a finish surface composed of gravel embedded in asphalt. How much gravel is normally used on a typical square foot of such a flat roof?

   A. 4 pounds   C. 100 pounds
   B. 12 pounds D. 350 pounds

Built-up roofs, such as that described in this question, are normally composed of a waterproof base sheet, several plies of roofing felt mopped between layers with asphalt, and a final flood coat of asphalt into which gravel or slag is embedded. This type of roof weighs about 600 pounds per square, a square being a roofing material measure equal to 100 square feet of roof surface. The weight of the gravel accounts for over half that total, or about 400 pounds per square, which is 4 pounds per square foot (correct answer A).

7. In selecting a roof covering, which of the following factors must one consider?

   I. Slope of roof
   II. Weight of roofing
   III. Water resistance of roofing
   IV. Fire resistance of roofing
   V. Climatic conditions

   A. I, II, III   **C.** I, II, IV, V
   B. III, IV, V   **D.** I, II, III, IV

The selection of roofing materials involves many considerations; however, one should not have to speculate about the water resistance of a material (III). By definition, roofing is the material used to cover a building, make it waterproof, and protect it from the natural elements. Therefore, all roof coverings inherently resist water. The other factors included in correct answer C all have a direct bearing on the choice of roofing. Slope and weight (I and II) may limit the use of several materials, and fire resistance (IV) determines the Underwriters Laboratories' (UL) classification of all roof coverings. Climate (V) refers not to rainfall as much as to wind, hail, atmospheric corrosion, etc., all of which affect one's final choice of roof covering material.

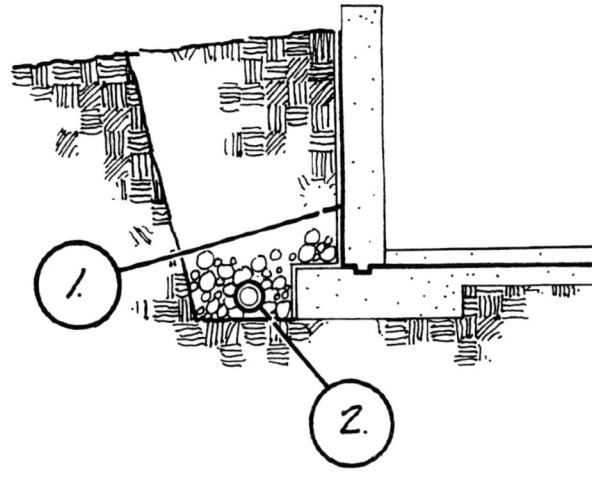

8. Shown above is a common method used to waterproof a concrete wall below grade. The surface indicated at point #1 is intended to

   A. function as a temporary concrete form.
   B. retain the soil prior to membrane application.
   C. provide a waterproof membrane.
   D. protect the waterproof membrane.

Concrete walls below grade are commonly waterproofed to resist ground water infiltration, usually under hydrostatic pressure. This is accomplished by means of a continuous membrane around the outside wall surfaces. The membrane consists of several plies of bituminous saturated felt, or cotton fabric, or a combination of these materials. It is sealed and held in place with asphalt or tar, which is hot-mopped, similar to the application of a built-up roof surface. Of particular concern is the danger of rupturing the membrane during the backfilling process. For this reason, an additional protective layer is applied over the membrane prior to backfilling (correct answer D). This layer consists of portland cement mortar, fiberboard, or sometimes a single wythe of masonry.

9. With reference to the illustration of the previous question, the device indicated at point #2 should be

  A. placed parallel to the footing in plan and level with the floor slab.
  B. located at the floor slab elevation, instead of as shown.
  C. covered with a vinyl sheet prior to placing the gravel backfill.
  D. at least six inches in diameter and laid with open joints.

The device shown is a drain tile, the purpose of which is to carry subsurface water away from basement walls and floors and into a storm sewer system. Drain tiles do run parallel to footings, but they must be sloped, rather than level (A) so that the water will run. The location of the tile is correct as shown (B), which is at least six inches below the floor slab. Prior to backfilling, drain tiles are often covered with wire screening or building paper to prevent clogging the drain with fine particles. Covering the tiles with a vinyl sheet (C) would prevent water from entering the drainage system and thus, defeat its purpose. Finally, with correct answer D we have an accurate statement; joints are left open so that subsurface water can enter the system and be carried away, eliminating the potential danger of excessive water undermining the footings.

10. Select the correct statement from among those that follow.

  A. Water-resistant, water-repellant, and waterproof are terms that may be used interchangeably to denote something that is impervious to water.
  B. Effective dampproofing generally involves continuous bituminous membranes.
  C. Concrete slabs in direct contact with moist ground will not absorb moisture unless the ground water is under pressure.
  D. Water-repellant admixtures for concrete decrease water absorption by reducing the capillary attraction of the voids in concrete.

Although all the statements pertain to moisture control, this is actually four questions in one, and candidates must verify at least three of the statements to find the correct answer. Except for statement D, which is the correct answer, each of the statements is in some way incorrect. The three terms about water permeability (A) have different meanings from one another. Beginning with water-resistant, each subsequent term indicates a higher level of water resistance. Concerning dampproofing (B), most often it involves the application of liquid coatings, not the use of membranes, which are used in waterproofing. Statement C is also false; an ordinary concrete slab in contact with moist earth, whether or not under pressure, will absorb moisture through capillary action. That is the reason that water-repellant admixtures, mentioned in statement D, are commonly used. Stearates or asphalt emulsions considerably reduce capillary action, if no hydrostatic pressure is involved. Incidentally, these admixtures also reduce the strength of the concrete and provide limited protection, not complete waterproofing.

PARKING

1. Select the correct combination of statements concerning parking lot efficiency.
   I. Rectangular parking lots are the most efficient.
   II. An efficient parking lot width is about 75 feet.
   III. Traffic circulation should be arranged along the perimeter of the parking lot.
   IV. Employee parking and customer parking should be separated.
   V. Traffic aisles should be double loaded.

   A. I, IV, V   **C.** I, II, IV
   B. II, III, V   **D.** I, III, IV

   Each of the statements is true with the exception of II and III. An efficient parking lot width (II) is one in which cars can be parked on both sides of a traffic aisle (double-loaded), regardless of the angle, and have no wasted space. (See the sketch below.)

   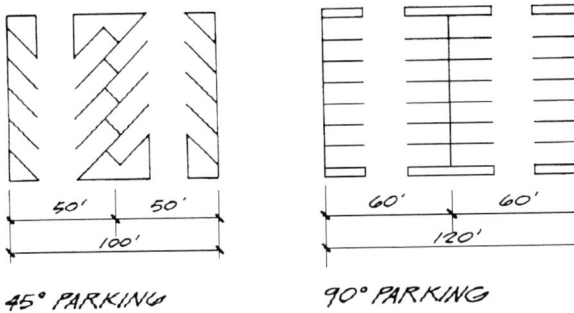

   For 45 degree angle parking, this dimension is about 50 feet, and for 90 degree angle parking, the dimension is about 60 feet. An inefficient parking area width is one between 60 and 100 feet. This range of parking widths should be avoided, since it is too wide for one double-loaded aisle and not wide enough for two such aisles. In statement I, rectangular lots are most efficient because the greatest number of rectangular parking stalls and driveways can be accommodated with the least amount of wasted space. Concerning statement III, the perimeter of a parking lot should be lined with parking stalls, not circulation, since this arrangement permits the maximum parking capacity, which is the most important consideration for optimum land usage. Finally, employee and customer parking should be separated (IV) so that the most convenient spaces are reserved for customers. This will also segregate areas of high and low turnover. The correct combination of answers is contained in choice A.

2. Vehicles entering a parking lot from the street generally have a different effective turning radius than vehicles exiting from the same lot. The reason for this is because
   A. drivers entering a parking lot are usually less familiar with the lot layout than those exiting.
   B. drivers entering a parking lot are usually moving at a faster rate than those exiting.
   C. drivers entering a parking lot usually take less care than those exiting.
   D. entrance driveways are usually wider than exit driveways, since it is assumed that those exiting have greater control than those entering.

   Vehicles entering a parking lot from the street are invariably traveling at a faster rate than those exiting (correct answer B), since they are in the process of slowing down from normal street speeds. Those exiting are moving along a restricted aisle width and are, therefore, traveling more slowly. In addition, vehicles exiting generally slow or even stop completely when they approach the street. This slower speed enables a driver to turn within a radius somewhat smaller than one traveling at a greater speed.

3. With reference to large open parking areas, all of the following are advantages of 90 degree parking, with the exception of
   A. it produces the highest car count.
   B. it allows for two-way traffic flow.
   C. it permits traffic aisle dead ends.
   D. it allows cars to either head into or back into the spaces.

All of the statements listed are advantages associated with 90 degree or perpendicular parking, with the exception of correct answer D. It is not that heading into or backing into is not unique to perpendicular parking, it is simply that this maneuver is not a particular advantage. In the great majority of cases, drivers who attempt to back in will obstruct the traffic flow for a longer period of time than those who head in. Additionally, in most large parking areas, concrete bumpers or wheel stops are installed for cars heading in, which places them about 36 inches back from the end of the space. For cars backing into a space, bumpers are set about 54 inches from the end of the space. This variation is caused by the difference between front and rear car overhangs. Consequently, cars that back in, rather than head in to such a space could possibly damage a wall or other car at the end of the space. Incidentally, traffic aisle dead ends (C) are desirable when strict traffic control is necessary, since they force cars to leave by the same traffic aisle they entered.

4. Match the following parking modes with the single descriptive word that most usually applies.
   I. Street parking
   II. Large parking lots
   III. Ramped parking structures

   1. unsightly
   2. expensive
   3. convenient

   A. I-2, II-1, III-3
   B. I-1, II-3, III-2
   C. I-3, II-2, III-1
   D. I-3, II-1, III-2

Depending on the particular way in which parking is provided, the descriptive words could conceivably apply to any of the parking modes listed. Street parking (I), for example, can be considered expensive, or at least an inefficient use of costly public streets, if it reduces traffic flow lanes and creates congestion. There are those who might even consider it unsightly. However, street parking is most generally thought of as convenient (3), since one has the chance of parking adjacent to one's destination. Large parking lots (II) may also be convenient, if no other parking exists, but most often their size makes them just the opposite. They are also considered inexpensive and unsightly (1) because there is rarely any relief from the monotony of paving and the expanse of cars. Finally, ramped structures, above or below grade, are sometimes convenient and occasionally unsightly, but they are always expensive. The correct combination, therefore, is found in answer D.

5. Select the *incorrect* statement concerning traffic flow in a parking area.
   A. Traffic aisles should be arranged so that they lead towards the buildings they serve.
   B. Angled parking requires one-way traffic flow.
   C. Perpendicular parking layouts usually lead to the most rapid traffic flow.
   D. Circulation of traffic within large parking areas should be continuous.

All of the statements are true with the exception of correct answer C. Angled parking nearly always leads to more rapid traffic flow than perpendicular parking, because one can pull in and out of an angled space more quickly; hence, traffic is obstructed or halted for a shorter period of time. Concerning choice A, traffic aisles generally lead toward buildings so that the buildings are more visible, and customers can walk along traffic aisles instead of between cars.

PLUMBING

1. Which of the following is not a backflow preventer?
   A. Double check valve
   B. Gate valve
   C. Vacuum breaker
   D. Air gap fitting

   A backflow preventer is a plumbing device installed between potable and nonpotable water systems, which allows flow from the potable to non-potable, but not in the opposite direction. It often consists of two check valves in series (A) that allow the flow of fluid in only one direction. When the flow is reversed, the check valves are designed to close, thus preventing backflow. A gate valve (B) is a manually-actuated shut-off valve, consisting of a sliding plate or gate that is perpendicular to the flow. Since it is not a backflow preventer, it is the correct answer to this question. A vacuum breaker (C) is an automatic valve that admits atmospheric air into a supply pipe upon a reverse in flow. This causes the flow to stop, preventing backflow and siphonage. Finally, an air gap fitting (D) is a device that separates two drain pipes with an unobstructed vertical air path to the atmosphere. This assures that backflow will not take place in the event a sudden pressure loss and momentary vacuum occur.

2. The purpose of a siamese connection is to
   A. drain a sprinkler system without causing disruption or damage.
   B. provide two exterior hose connections, one for the fire department and the other for building maintenance.
   C. provide a means of switching the domestic water system to a fire sprinkler system.
   D. allow the fire department to connect two supply hoses to a building's standpipe system.

   A siamese connection is the common term applied to the duplex hose fitting to which fire department hoses can be attached (correct answer D). The fitting has threads that match fire hoses, and upon connection to fire department pumpers, water can be pumped into the standpipe or fire sprinkler system.

   The following data applies to questions 3 and 4:

   A 20-story building, with 13-foot story heights, requires 30 psi water pressure at the roof for a wet standpipe outlet. The municipal water main in the street is five feet below street level, and it has a minimum pressure of 90 psi. Pressure losses through the meter and piping system are assumed to be 20 psi.

3. Without employing a booster pump, the municipal water pressure will be sufficient to service all floors up to and including the
   A. 5th floor.
   B. 7th floor.
   C. 10th floor.
   D. 11th floor.

4. Employing a booster pump in a pressurized system, the pump would have to be sized at
   A. 13 feet of head.
   B. 70 feet of head.
   C. 89 feet of head.
   D. 169 feet of head.

   The water pressure in the street of 90 psi must be reduced by the system losses of 20 psi, as well as the 30 psi pressure that is required. Therefore 90 − 20 − 30 = 40 psi of pressure that remains. Since 1 psi is equivalent to 2.3 feet of head, 40 psi × 2.3 feet = 92 feet of vertical height is all that is available. Converting this to stories of building,

92 feet ÷ 13 feet per story = approximately 7 stories (correct answer B).

In question 4, the booster pump would be required to accommodate those stories of the building above the seventh story that are not handled by the street pressure described above. Therefore 20 stories − 7 stories = 13 stories × 13 feet per story = 169 feet (correct answer D).

5. Which among the following piping materials would serve satisfactorily for a domestic water system?
   I. copper
   II. lead
   III. cement-lined cast iron
   IV. vitrified clay
   V. plastic
   VI. black steel
   VII. galvanized steel

   A. I, III, V, VII
   B. I, II, III, IV
   C. I, V, VI, VII
   D. II, III, V, VII

Domestic water piping, which carries potable or drinking water, must be, most importantly, non-toxic. It must also be able to withstand high pressure, it must be easily connected with leak-proof joints, and it must resist corrosion. Among the piping materials listed that do not meet these requirements are lead, vitrified clay, and black steel. Lead (II) is not only highly toxic, but it cannot withstand high pressure. Vitrified clay (IV) piping has joints that are difficult to seal and which tend to leak under high pressure. The deficiency of black steel (VI) is that it is easily corroded. The correct combination of choices is found in correct answer A.

6. Traps are provided in plumbing fixture drain lines in order to
   A. intercept solid objects that might impede water flow.
   B. maintain an unobstructed vent.
   C. prevent sewer gases from entering a building.
   D. reduce the noise of water flow through the system.

A trap is a water seal fitting, often U-shaped, which is located in the drain line after the fixture. As it is filled with water, it forms a trap that prevents sewer gases and odors from reversing their flow and entering a building (correct answer C). Special traps do exist for the purpose of catching solids (A), but these are exceptional cases, such as dentists' sinks, which are designed to recover gold and silver particles. Traps have no effect on vents (B); vents maintain atmospheric pressure in the drain so that water from the trap will not be siphoned. Finally, traps have little to do with creating or preventing plumbing noises (D).

7. There are many impurities that can exist in a potable water system. Which of those following can best be removed with the use of a zeolite water softener?
   I. Turbidity
   II. Calcium ions
   III. Low pH
   IV. Organic matter

   A. I, II, III
   B. II, III
   C. II only
   D. IV only

Zeolite is a natural or synthetic resin that readily exchanges sodium ions for magnesium or calcium ions (II) that are found in hard water. For this reason zeolite is widely used for water softening. Turbidity (I), which is caused by sediment, is best removed from water by filtration. Low pH (III) is acidity, and this is best treated by the addition of an alkaline substance to the water. Finally, organic matter (IV), which actually comprises living organisms in the water, requires chlorination treatment. The correct answer is C.

SOCIOLOGY

1. To discourage vandalism in a housing project, which of the following details should be provided at exterior spaces?

   I. Exterior paths and entrance doors arranged for visibility.
   II. Exterior paths arranged in curves, rather than rigid straight lines.
   III. Exterior lighting used along paths and at entrances.
   IV. Natural and attractive exterior materials used to inspire respect.

   A. I and IV
   B. I and III
   C. I, II, and III
   D. II, III, and IV

   Vandalism, which is the willful destruction of property, is an unfortunate fact of modern life, and a subject with which architects must often be concerned. To deal with the perils of vandalism, one must understand the techniques of security. For example, at exterior spaces the key technique is surveillance, with local residents taking the responsibility over their own locality. In this regard one would arrange paths for high visibility (I), and provide well-lighted grounds (III). On the other hand, one would avoid curved paths (II) because they tend to obscure the view ahead. Finally, one must be realistic and not assume that attractive or natural materials are immune to vandalism, for until social peace exists, architects must design defensively. The correct choices are found in answer B.

2. Select the correct statement concerning the concept of density.

   A. Density is approximately the same as crowding.
   B. Most means used to measure density involve calculations of area.
   C. High density invariably leads to a loss of environmental quality.
   D. A direct relationship exists between poverty and density.

   In the field of planning, the concept of density has often been misunderstood. To begin with, density is defined as the intensity of human activity occurring per unit of area. Therefore, the means used to measure density all rely on area (correct answer B), such as floor area ratio (the ratio of gross floor area of a building to its ground area), rooms per acre, or families per acre. Density must not be confused with crowding (A), since crowding involves perceptions based on a person's activity. For example, a theater has very high density, yet it does not produce a crowded feeling to each person in his own seat. Density, therefore, must not be considered as necessarily harmful (C and D), since at times, it may very well be a desirable quality leading to diversity and interest.

3. Research on environmental behavior indicates all of the following *except*

   A. large, public open spaces create a sense of community.
   B. the layout of commercial stores affects shopping behavior.
   C. work efficiency is affected by both temperature and quality of light.
   D. design decisions often influence interpersonal relationships.

Understanding how people use and react to the spatial environment is the key to creating humane designs. In this regard, recent research has provided architects with a great deal of valuable information that can be applied during the planning process. For example, all of the statements in this question are true and useful facts, with the exception of correct answer A. The public square, or plaza, is a popular planning feature, but quite often it receives little public use. Historically, this kind of space was necessary for political, religious, or commercial activities. Today, however, our activities differ considerably from the past, and hence, the public plaza may actually create a void, rather than an important social setting.

4. If a designer wanted to impart symbolic importance to a public institutional building, the form of the structure might include

   I. a number of wide entrance doors.
   II. a symmetrical arrangement.
   III. a long flight of entrance steps.
   IV. small, randomly placed windows.
   V. a flagpole.

   A. I, II, V   **C.** II, III, V
   B. I, III, V   **D.** II, III, IV

   Architectural symbolism has existed and been employed for thousands of years. Throughout nearly all that time, informal or casual arrangements have signified relative unimportance, while importance has been characterized by rigid formality. In the case of an important public building, therefore, the designer would include a symmetrical arrangement (II), which is regarded as formal, and a long flight of entrance steps (III), which accentuates the insignificance of the user, and thus, the relative importance of the institution. Additionally, one would use a flagpole (V), since this object symbolizes the institution's important association with the governing state or country. The use of many entrance doors (I) actually diminishes the importance of a single principal entrance; furthermore, a tall, rather than wide door would signify greater prominence. Finally, for the reasons already mentioned, one would avoid random placement of openings (IV). The correct combination is found in answer C.

5. It is common practice today to preserve and restore older buildings that may have outlived their original function. Following are several statements which justify preservation. Which of these has the least validity?

   A. Preservation of older buildings provides a cultural and often historical link with the past.

   B. Restoration of older buildings often provides valuable enclosed space at a cost comparable with or lower than new construction.

   C. Preservation of buildings that are familiar to residents of a neighborhood usually generates a spirit of cooperation by these residents.

   D. Modern concrete, steel, or glass structures that replace older buildings tend to reduce the quality of human scale, warmth, and interest in the community.

   The preservation and restoration of older buildings has become a commendable activity for many of the reasons listed (A, B, and C). It is invalid, however, to assume that modern buildings cannot also possess scale, warmth, or interest, as stated in correct answer D. Communities that habitually replace older buildings may tend to lose touch with their heritage, but that does not mean that they are devoid of desirable human qualities.

SOLAR ENERGY

1. The principal difference between passive and active solar systems is that

   A. passive is usually retrofitted into an existing structure, while active is generally provided in new construction.

   B. passive employs the energy of the sun, while active relies on imported energy, such as electricity.

   C. passive involves natural energy flows, while active involves mechanical energy flows.

   D. passive is used primarily to heat, while active can both heat and cool.

The distinction between passive and active solar systems is something all candidates should know. Essentially, passive systems collect and transmit solar heat by completely natural means, such as radiation, conduction, and natural convection. Active systems, on the other hand, employ mechanical equipment to collect and transmit solar heat, such as flat plate collectors, fans, pumps, etc. (correct answer C). Passive systems operate on the energy available in the immediate environment, whereas active systems rely on imported energy, such as electricity, to power fans and pumps that are necessary for the system to operate. In that limited sense, statement B is not entirely false. However, one must remember that both systems are solar; they both rely on the radiant energy of the sun.

2. In many desert areas of the country, the climate during the summer is hot and dry during the day, but cooler and quite comfortable at night. In such an area, which of the following building conditions would be desirable?

   I. Exterior shading devices
   II. Dark color on the south facade
   III. Light colored roofing
   IV. Insulation on the exterior of masonry walls
   V. Vented openings facing the prevailing wind

   A. I, III, V
   B. II, III, IV
   C. I, II, IV, V
   D. I, II, III, IV, V

Candidates should be aware that an understanding of solar energy includes consideration of cooling, as well as heating. In this question we are dealing with a situation common to a large segment of our country, and as correct answer D states, all the conditions listed are desirable. Items I, III, and V are conditions which should be obvious to all designers. Items II and IV, however, may require an explanation. By using a dark color on the south facing wall, low south sunlight will be absorbed in the winter, when it is required, while creating no serious problem in the summer, when the sun is high. The use of thermal insulation on the exterior wall insures that only a small portion of the exterior heat is conducted through the building's skin to the interior. The insulation, of course, must be protected from the weather and damage by applying plaster or siding over it. Exterior insulation also has the advantage of retaining stored heat within the wall during the winter months.

3. With regard to a roof pond passive solar system, select the correct statement from among those which follow.

   A. In a roof pond system a thermal mass is not required.
   B. A roof pond system provides efficient cooling, but inadequate heating.
   C. A roof pond system provides efficient heating, but inadequate cooling.
   D. Roof pond systems require an exterior cover to control radiation.

   All passive solar systems require a means of collecting solar heat, as well as a thermal mass to absorb, store, and later distribute this heat throughout a building. A roof pond system is no exception; it employs water for its thermal mass (A), usually in the form of water-filled transparent plastic bags. This system is equally suited to both heating in winter and cooling in the summer (B and C). In winter the pond is exposed to sunlight during the day and then covered by insulating panels at night (correct answer D). Heat that is collected and stored in the water-filled bags radiates through the ceiling directly to the space below. In summer the cycle is reversed: the pond is covered during the day, to protect it from the sun, and uncovered at night to allow the pond to be cooled by natural convection and radiation to the cool night air. By morning, the cool pond is ready to absorb heat from the space below.

4. Some passive solar systems employ a conventional-type greenhouse that is attached to a structure. Which of the following statements about this system is *incorrect*?

   A. This system also employs a thermal storage wall.
   B. This system is virtually the same as a thermal storage wall system.
   C. This system is easily added to the south wall of an existing building.
   D. This system allows one to grow fresh vegetables throughout the year.

   An attached greenhouse solar system consists of a conventional greenhouse that is readily constructed on the south side of a structure (C), with a thermal mass storage wall separating the greenhouse from the rest of the building (A). It is essentially a combination of direct and indirect gain systems; that is, the greenhouse space is directly heated by sunlight, the back wall absorbs the heat, and a portion of this heat is transferred into the building. The system differs from a thermal storage wall system (incorrect statement and correct answer B), which is an indirect gain system. In such systems, sunlight strikes the thermal mass wall where it is absorbed by the mass, converted to thermal energy (heat), and then transferred to the rest of the building. Statement D expresses a unique advantage of the attached greenhouse solar system. The greenhouse area functions efficiently as a space in which plants can thrive throughout the year in almost any part of the country.

5. If a building were to be planned with regard to the sun's impact on space heating, the most desirable forms of the building would be which of the following?

I. Circular
II. Square
III. Elongated on the north-south axis
IV. Staggered vertically or horizontally
V. Stacked vertically

A. I, III
B. IV, V
C. I, II, IV
D. II, III, IV, V

Buildings that disregard the sun's impact frequently require large amounts of energy to heat and cool. When determining a building's shape, designers should consider the solar orientation that will minimize the use of energy. In general, the optimum shape of a building is one which loses a minimum amount of heat in the winter and gains a minimum amount of heat in the summer. Considering the choices of this question, a circular or square shape (I and II) are not desirable, since they have the least amount of enclosing walls for the area. This means that sunlight will strike an absolute minimum amount of exterior surface. The very best shape is a form elongated in the east-west direction (not the north-south as stated in III), as this form will derive maximum heat gain during the winter months, while exposing the shorter east and west sides to minimum heat gain in the summer. Staggered or stacked buildings (IV and V) can also be beneficial, since direct sunlight will strike large surface areas, especially if the building's longer dimensions are oriented to the south. The correct combination of choices is found in correct answer B.

STRUCTURES

Note: Questions 1 through 7 require the candidate to perform calculations. Questions of this type are therefore given extra weight on the actual exam.

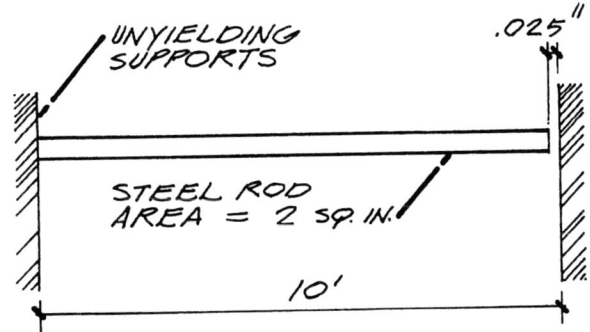

1. What is the temperature increase necessary to close the gap?
   A. 38.46°F   C. 16.03°F
   B. 32.05°F   D. 64.10°F

All materials expand when heated and contract when cooled. The ratio of unit strain to temperature change is called the coefficient of thermal expansion and is constant for a given material. For steel, its value is 0.0000065.
Coefficient of thermal expansion
$\alpha = 0.0000065$
Change of length $\Delta = 0.025$ in.
Length L = 10 ft. = 120 in.
Temperature change (°F) = dT

$$\alpha = \frac{\Delta/L}{dT} \quad dT = \Delta/\alpha L$$

$dT = 0.025/(0.0000065)(120)$
$= 32.05°F$ *(correct answer B).*
Note that the length and the change of length must be in the same units. In this case, it was necessary to convert the 10 foot length to 120 inches.

2. What tensile force is required to close the gap of the previous question, if there is no temperature change?
   A. 1,250#   C. 12,500#
   B. 6,250#   D. 15,000#

In problems involving deformation, or change in length, it is important to remember that the modulus of elasticity (E) is equal to the ratio of unit stress (P/A) to unit strain (Δ/L).

$$E = \frac{P/A}{\Delta/A} = \frac{PL}{A\Delta} \quad \text{or} \quad P = \frac{EA\Delta}{L}$$

Since E for steel is 30,000,000

psi, we have $P = \frac{(30,000,000)(2)(0.25)}{120}$

$= 12,500\#$ *(correct answer C).*
Note that the units must be consistent. E is in pounds per square inch, P in pounds, A in square inches, Δ in inches, and L in inches.

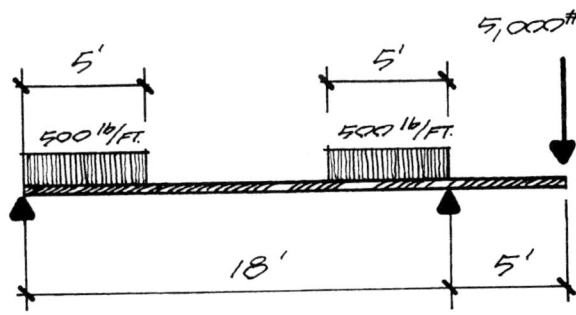

3. What is the maximum moment in the beam above?
   A. 25,000'# C. 1,235'#
   B. 11,250'# D. 50,000'#

In any beam problem, we must first determine the reactions, then the shears, and finally the moments. To solve for the left reaction, we take moments about the right support.

$\Sigma M = 0$
$5000\,(5) - 500\,(5)(2.5) - 500\,(5)(13 + 2.5) + R_L\,(18) = 0$
$25{,}000 - 6{,}250 - 38{,}750 + 18R_L = 0$
$18R_L = 38{,}750 + 6{,}250 - 25{,}000 = 20{,}000$
$R_L = 20{,}000/18 = 1111.1\#$

Now solving for the right reaction,

$\Sigma V = 0$
$1{,}111.1 + R_R - 500\,(5) - 500\,(5) - 5{,}000 = 0$
$R_R = 2{,}500 + 2{,}500 + 5{,}000 - 1{,}111.1$
$R_R = 8{,}888.9\#$

Although not absolutely necessary, we will now draw the shear diagram for this beam.

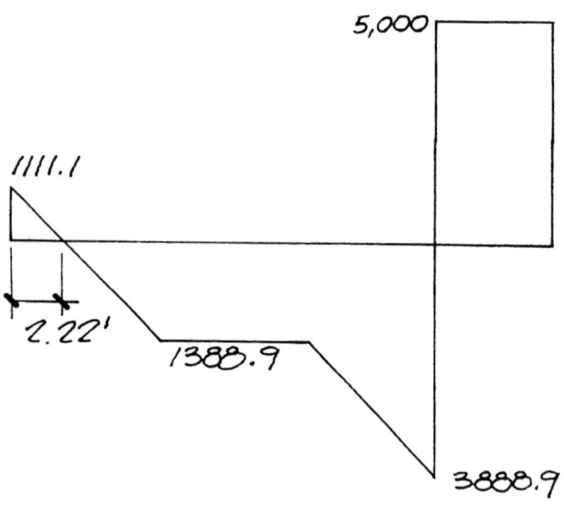

The maximum moment occurs where the shear is equal to zero, which occurs at two places: 2.22 feet from the left support and at the right support. We therefore calculate the moment at each of these two points. Moment at 2.22' from left support = $1{,}111.1\,(2.22) - 500\,(2.22)^2/2 = 1{,}235\,'\#$
Moment at right support = $5{,}000(5) = 25{,}000\,'\#$
This is the maximum moment (correct answer A). Incidentally, this moment is negative—that is, it causes tension in the top beam fibers and compression in the bottom fibers.

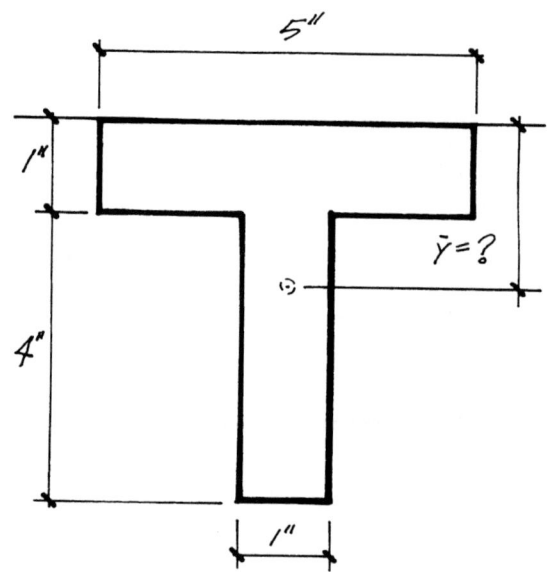

4. Find the distance $\bar{y}$ to the centroid of the area shown above. $\bar{y} =$

A. 1.61" C. 1.75"
B. 3.39" D. 1.89"

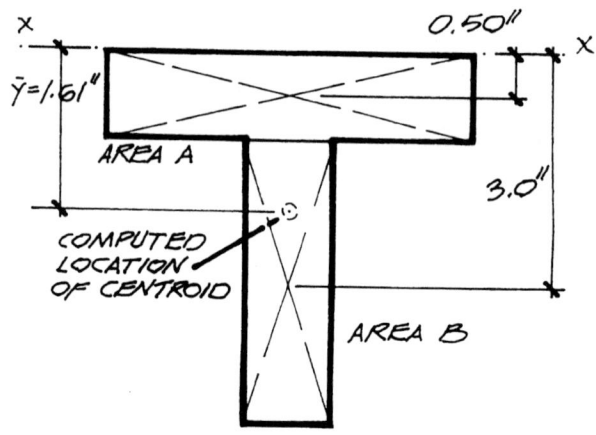

The area is divided into two parts, A and B, as shown above. Calculate the area of each part and the statical moment of each part about axis x-x, the top of the area.

Part	Area	y	Area xy
A	5 × 1 = 5.0	0.50	5.0 × 0.50 = 2.5
B	4 × 1 = 4.0	3.0	4.0 × 3.0 = 12.0
Total		9.0	14.5

$\bar{y}$ = *the distance from axis x-x to the centroid of the area* = 14.5/9.0 = 1.61″ *(correct answer A).*

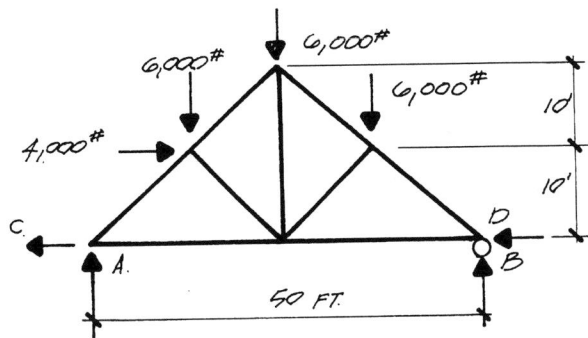

5. What are the reactions A, B, C, and D on the roof truss shown above supporting the vertical and horizontal loads shown. The truss is free to roll at the right support only.

   A. A = 9,800#, B = 8,200#, C = 4,000#, D = 0
   B. A = 8,200#, B = 9,800#, C = 4,000#, D = 0
   C. A = 8,200#, B = 9,800#, C = 0, D = 4,000#
   D. A = 9,000#, B = 9,000#, C = 4,000#, D = 0

This problem involves basic statics, and the methods used in its solution are fundamental. If you have difficulty solving the problem or following our solution, you should spend time reviewing statics. Since the truss is free to roll at the right support, no horizontal reaction is possible there. Therefore, reaction D is zero.

$\Sigma H = 0$
4,000 − C − D = 0
4,000 − C − 0 = 0
C = 4,000#

To solve for reaction B, take moments about the left support.

$\Sigma M = 0$
6,000 (50/4) + 6,000 (50/2)
+ 6,000 (3 × 50/4) + 4,000 (10) − 50B = 0
50B = 6,000 (12.5 + 25 + 37.5) + 40,000
B = (450,000 + 40,000)/50
B = 9,800#
$\Sigma V = 0$
A + B − 6,000 − 6,000 − 6,000 = 0
A + B = 18,000
A + 9,800 = 18,000
A = 18,000 − 9,800 = 8,200#
Summarizing, A = 8,200#, B = 9,800#, C = 4,000#, D = 0
The correct answer is B.

6. What is the magnitude and direction of the resultant of the three forces shown?

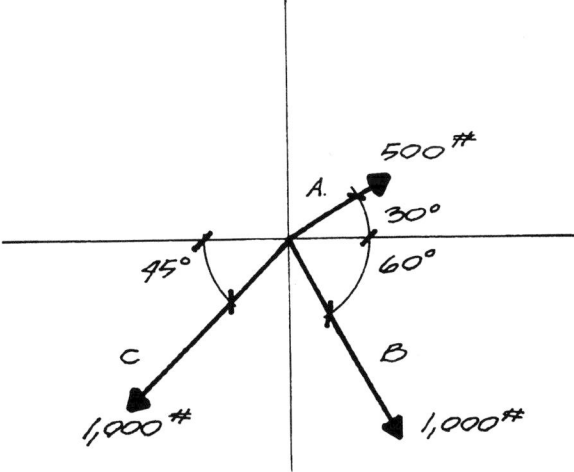

   A. 1,000.0# at 60.0° to the horizontal, downward to the right.
   B. 1,140.9# at 87.85° to the horizontal, downward to the right.
   C. 875.5# at 62.16° to the horizontal, downward to the right.
   D. 1,342.2# at 80.31° to the horizontal, downward to the right.

This problem can be solved either analytically or graphically. We will go through the analytical solution. Set up a table with each force and its horizontal and vertical components.

Force	Value	Horizontal Component	Vertical Component
A	500#	500 cos 30° = 433.0	500 sin 30° = 250.0
B	1,000#	1,000 cos 60° = 500.0	−1,000 sin 60° = −866.0
C	1,000#	−1,000 cos 45° = −707.1	−1,000 sin 45° = −707.1
Total		ΣH = 225.9	ΣV = −1323.1

The value of the resultant =

$$\sqrt{(\Sigma H)^2 + (\Sigma V)^2}$$

$$\sqrt{(225.9)^2 + (-1323.1)^2} = 1,342.2\#$$

The angle θ between the resultant and the horizontal is such that

$$\tan\theta = \frac{\Sigma V}{\Sigma H} = \frac{-1,323.1}{225.9} = -5.86$$

θ = 80.31° with the horizontal.

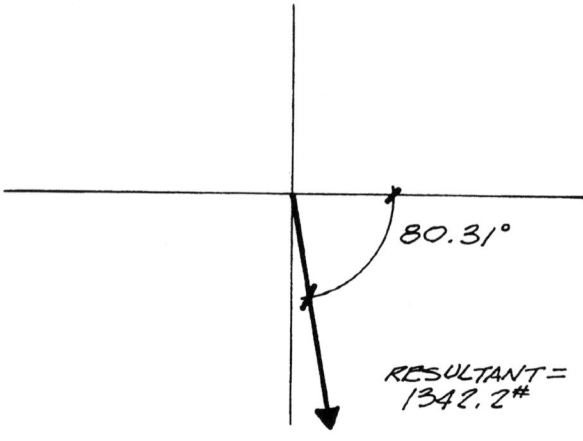

Note that the horizontal components of forces A and B are positive, since they act towards the right; and the horizontal component of force C is negative, since it acts towards the left. Also, the vertical component of force A is positive, since it acts upward; while the vertical components of forces B and C are negative, since they act downward. Since ΣH is positive and ΣV is negative, the resultant acts downward and to the right, as shown.
The correct answer is D.

7. A uniform ladder 18 feet long and weighing 60 pounds leans against a smooth vertical wall at an angle of 70° with the ground. What force does the ladder exert against the wall?

A. zero C. 21.84#
B. 82.42# D. 10.92#

Although this problem does not directly relate to the design of buildings, it does test a candidate's understanding of basic statics, which has always comprised an important part of the structural exam.

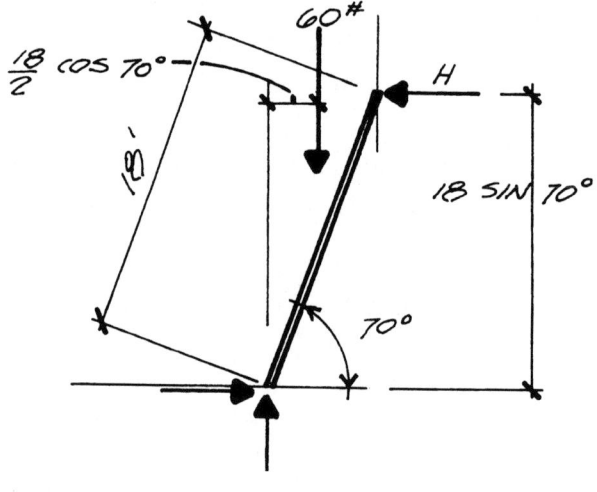

The ladder and the forces acting on it are shown on the previous page. H represents the reaction by the wall on the ladder, which is equal and opposite to the push of the ladder against the wall. H acts perpendicular to the wall, and hence horizontally, because the wall is smooth (frictionless).

We take moments about the foot of the ladder to solve for reaction H.

$\sum M = 0$

$60\left(\dfrac{18}{2} \cos 70°\right) - H(18 \sin 70°) = 0$

$H = 60(9 \cos 70°)/18 \sin 70° = 10.92\text{\#}$ *(correct answer D).*

8. The symbol 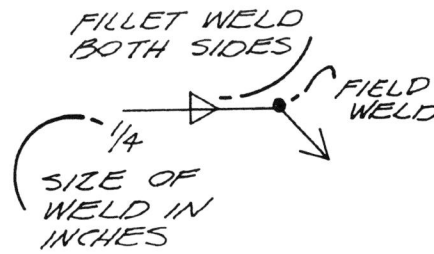 refers to

   A. $\dfrac{1"}{4}$ fillet weld both sides (field).

   B. $\dfrac{1"}{4}$ fillet weld both sides (shop).

   C. $\dfrac{1"}{4}$ groove weld both sides (field).

   D. $\dfrac{1"}{4}$ groove weld both sides (shop).

Each part of the welding symbol has a meaning, as indicated below.

The correct answer is therefore A. Standardized welding symbols quickly indicate to all concerned the exact welding detail to be used for each joint or connection, and candidates should understand the meaning of these symbols.

9. Why is compression steel used in reinforced concrete beams?

   A. To reduce initial deflection

   B. To reduce the amount of tensile steel

   C. To increase the load-carrying capacity of shallow beams

   D. To reduce the number of stirrups

Sometimes a reinforced concrete beam is too shallow to resist the applied moment. In that case, reinforcing steel may be embedded in the compression face of the beam to increase its moment capacity (correct answer C). The use of compressive reinforcement reduces the long-term deflection of the beam, but not its initial deflection (choice A). Compressive reinforcement has little effect on the amount of tensile steel required (B) and no effect on the number of stirrups needed (D).

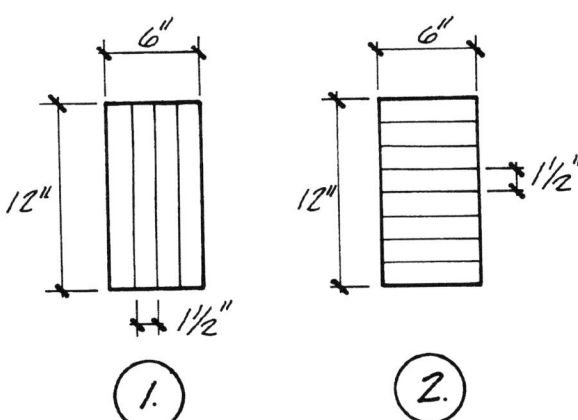

10. Shown above are two methods for building up a wood beam by nailing individual laminations together. Which of the following are the advantages of beam 1 over beam 2?

   I. Less deflection

   II. Lower flexural stress

   III. Lower horizontal shear stress

   IV. Higher allowable flexural stress

   A. I, II, III, IV
   B. I, II, III
   C. I, II
   D. III only

This question tests your understanding of basic beam theory. Beam 1 consists of 4 beams 1 1/2" wide by 12" deep. The section modulus of each is $bd^2/6 = 1.5(12)^2/6 = 36$ in^3, and the total section modulus is equal to $4 \times 36 = 144$ in^3. The moment of inertia of each lamination is $bd^3/12 = 1.5(12)^3/12 = 216$ in^4, and the total moment of inertia is $216 \times 4 = 864$ in^4. Beam 2 consists of 8 beams 6" wide by 1 1/2" deep. The section modulus of each is $bd^2/6 = 6(1.5)^2/6 = 2.25$ in^3, and the total section modulus is $8 \times 2.25 = 18$ in^3. The moment of inertia of each lamination is $bd^3/12 = 6(1.5)^3/12 = 1.6875$ in^4, and the total moment of inertia is equal to $8 \times 1.6875 = 13.5$ in^4. Summarizing, S for beam 1 is 144 and S for beam 2 is 18. I for beam 1 is 864 and I for beam 2 is 13.5. Since the bending resistance of a beam is a function of its section modulus, beam 1 is 8 times as strong in bending as beam 2 ($144 \div 18 = 8$). Put another way, for the same span and load, the flexural stress of beam 1 will be only 1/8 that of beam 2. The deflection resistance, or stiffness, of a beam is a function of its moment of inertia. Therefore, beam 1 is 64 times as stiff as beam 2 ($864 \div 13.5 = 64$). Or, in other words, for the same span and load, beam 1 will only deflect 1/64 as much as beam 2. Choices I and II are therefore correct. The horizontal shear stress for a rectangular beam is equal to $\frac{3}{2}\frac{V}{bd}$. For beam 1, bd is equal to $4 \times 1.5 \times 12 = 72$ square inches. For beam 2, bd is $8 \times 6 \times 1.5 = 72$ square inches. Therefore, the horizontal shear stresses in beams 1 and 2 are equal, making choice III incorrect. Choice IV is also incorrect, because there is nothing in the problem statement which relates to allowable stress. C is therefore the correct answer. It is not necessary to actually perform any calculations in this question—we have done so in order to show that narrow deep beams ▯ have much greater bending resistance and resistance to deflection than wide shallow beams ▭.

Also, that horizontal shear stress in a rectangular beam is a function only of its area. Also, note that the individual laminations are nailed together. If they were bonded by glue, or otherwise connected so as to develop adequate shear strength, they could be made to act as one beam instead of separate laminations.

11. Which of the following best describes posttensioned construction?

   A. A prestressed concrete system in which prestressing strands are tensioned between abutments prior to placing concrete in the beam forms

   B. A prestressed concrete system in which prestressing wires are placed in a hollow sleeve in the beam and tensioned after the concrete is placed and has acquired its strength

   C. A floor system supported by cables, which are tensioned after pouring the floor slab

   D. A composite system consisting of a concrete slab placed over steel beams, in which the shoring is not removed until the concrete has attained its specified strength

There are two basic methods by which concrete is prestressed: pretensioning and posttensioning. In the pretension method, the steel is tensioned before placing the concrete, as described in choice A. Posttensioning involves tensioning the steel after the concrete is placed (correct choice B). Both C and D are irrelevant answers.

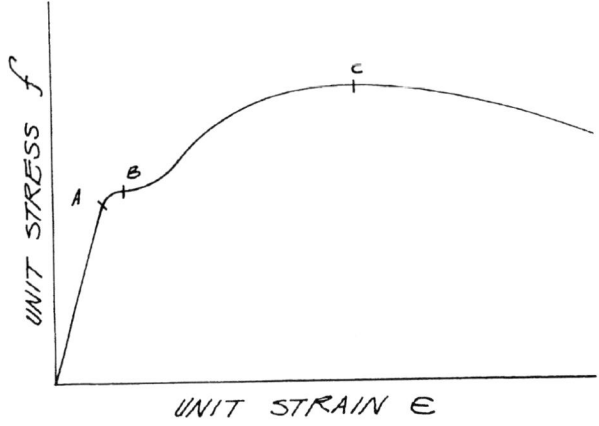

12. In the stress-strain curve shown, points A, B, and C are called, respectively, the_____, _____, and_____, and the slope f/ε is the _____.

- **A.** elastic limit, yield point, ultimate strength, modulus of elasticity.
- **B.** yield point, elastic limit, ultimate strength, modulus of elasticity.
- **C.** elastic limit, yield point, point of rupture, moment of inertia.
- **D.** working stress, yield point, ultimate strength, factor of safety.

If a member, such as a steel bar, is subject to an axial load that gradually increases from zero until it breaks, and if its unit stress (load divided by area) is plotted against its unit strain (deformation divided by length), the resulting curve will have approximately the shape shown to the left. In this curve, point A is called the elastic limit and represents the maximum unit stress that can be developed in the material without causing a permanent set. At any stress greater than the elastic limit, the member will not return to its original length when the load is reduced to zero. As the load on the member is increased, causing a stress greater than the elastic limit, a stress is reached at which the material continues to deform without an increase in load. That stress, B on the curve, is called the yield point. If the load on the member increases still further, the maximum unit stress that can be developed in the member is reached. That stress is called the ultimate strength and is represented by point C on the stress-strain curve. The numerical value of the ratio of unit stress to unit strain within the elastic limit, which is the slope f/ε of the straight-line portion of the curve, is called the modulus of elasticity of the material, whose symbol is E. The terms are correctly identified in answer A.

13. Select the *incorrect* statement concerning open web steel joists.

- **A.** An 18J5 joist usually has the same depth and configuration as an 18H5 joist.
- **B.** Steel used for J-series joists has a yield strength of 36,000 psi, while chord sections of H-series joists have a yield strength of 50,000 psi.
- **C.** Top chord extended ends can be used with H-series joists, but not with J-series joists.
- **D.** H-series and J-series joists may use either hot-rolled or cold-formed chords.

Open web steel joists are open web parallel chord members that are completely stan-

dardized as to length, depth, and carrying capacity, and are suitable for the support of both floors and roofs. They may utilize either hot-rolled or cold-formed steel (correct statement D). H- and J-series joists are virtually identical, except that the H-series joists are comprised of higher-strength steel (correct statements A and B), and therefore have greater load-carrying capacity. Both H- and J-series joists are often provided with top chord extended ends, which are used for short cantilevers (C). C is the incorrect statement and hence the answer to this question.

14. Shown in the next column is a section through a reinforced concrete slab and beam floor system. Which of the locations shown is best for a construction joint?

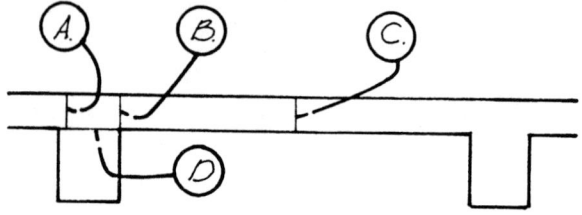

Reinforced concrete beams are generally constructed monolithically with the slab, and a portion of the slab acts with the beam to form a T-beam. Joints located near the slab-beam intersection are therefore not advisable. This rules out locations A, B, and D, leaving C as the only possible answer. C is also near the midspan of the slab, where the shear in the slab is minimum.

15. Select the correct statements. Punching shear is
    I. the tendency of a column to punch through its supporting footing.
    II. two-way shear.
    III. the same as beam shear.
    IV. investigated at a distance d from the face of the column.
    V. investigated at a distance d/2 from the face of the column.

    A. I, III, V C. I, II, IV
    B. III, IV D. I, II, V

A column supported by a footing tends to punch through the footing because of shear stresses which act in the footing around the perimeter of the column (I). The shear is a two-way action (II), and is calculated through the footing around the column on a perimeter a distance d/2 from the faces of the column (V). The other type of shear that occurs in footings is one-way or beam shear, which is calculated at a distance d from the face of the column. However, this is not a punching shear (incorrect statements III and IV). The correct statements are I, II, and V, as found in correct answer D.

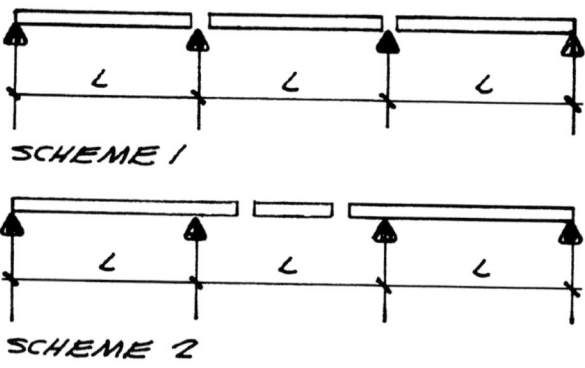

SCHEME 1

SCHEME 2

16. Two different schemes for a three-span beam are shown. Scheme 1 consists of three single-span beams. In Scheme 2, the outer beams overhang the inner supports and support the ends of the middle beam. Select the correct statements.

   I. Scheme 1 has greater positive moment than Scheme 2.
   II. Scheme 1 has greater negative moment than Scheme 2.
   III. Scheme 1 has greater deflection in the outer spans than Scheme 2.
   IV. Although Scheme 2 may utilize material more efficiently than Scheme 1, the cost of the special connections required generally precludes its use.

   A. I, II, and III   C. I and IV
   B. I and III   D. IV only

In wood and structural steel construction, simple beams are often employed, as in Scheme 1. However, various versions of Scheme 2 are also used. The overhanging action of the beams in Scheme 2 reduces the positive moment (I) and the end span deflection (III). There is no negative moment at all in Scheme 1, while in Scheme 2, negative moment occurs over the inner supports (II is incorrect). Finally, IV is also incorrect; it is true that Scheme 2 may utilize material more efficiently than Scheme 1. However, the connections required are seldom so complex or expensive as to preclude the use of Scheme 2. Since only I and III are correct statements, B is the right answer.

17. The compressive strength of concrete is based on
   A. standard cylinders tested at 28 days of age.
   B. standard cylinders tested at 7 days of age.
   C. tests of cores drilled from the completed concrete work.
   D. load tests of the completed concrete work.

Unless specified otherwise, the compressive strength of concrete is always based on standard cylinders tested at 28 days of age (correct answer A). Incidentally, the 7-day strength of concrete is approximately 60 percent of its 28-day strength. Core tests (C) and load tests (D) are only performed when cylinder tests indicate inadequate concrete strength.

18. Which of the shapes shown below is ideal for a column?

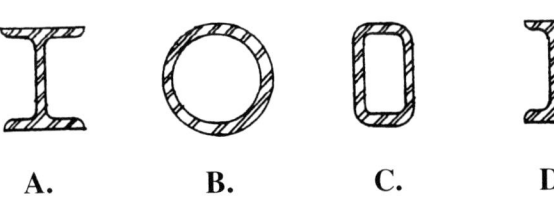

A.   B.   C.   D.

In theory, an ideal column is one for which the buckling tendency is equal in both directions. This occurs when the radius of gyration r in the x-x direction is equal to r in the y-y direction. This also means that the moment of inertia I is equal in both directions. These conditions are met by pipe columns (correct choice B) and square steel tubes. In actual practice, pipe columns and square steel tubes are not widely used because of the difficulty of making adequate beam connections.

19. In order to identify high-strength bolts which conform to ASTM A325, certain standardized markings are used. In this regard, which of the following is *incorrect*?

A. The bolt head is identified by the legend "A325".
B. The bolt head may be marked with three radial lines 120 degrees apart.
C. The standard nut marking consists of three circumferential marks spaced 120 degrees apart.
D. The nut may be marked with the letters "HS."

Questions of this type, testing a candidate's understanding of standard symbols used in construction, have appeared on past exams. It is therefore advisable to become familiar with such symbols. Heavy hex structural bolts which conform to ASTM A325 are identified on the top of the head by the legend "A325" and the manufacturer's symbol (A). At the option of the manufacturer, the bolts may be marked with three radial lines 120 degrees apart (B). The standard nut marking is as described in statement C. Only D is an incorrect statement, making it the answer to this question.

20. Which of the following statements is correct concerning two concentrically loaded reinforced concrete columns, one tied and one with spiral reinforcement but otherwise identical?

A. The spiral column can support more load.
B. The tied column can support more load.
C. Since the load is resisted by the concrete and longitudinal reinforcement only, both columns can support the same load.
D. For concrete strengths up to 4,000 psi, the spiral column can support more load; above 4,000 psi, the tied column can support more load.

Two types of reinforced concrete columns are in general use: tied columns and spiral columns. Tied columns are usually square or rectangular in cross-section and contain longitudinal reinforcing bars with separate lateral ties. Spiral columns are square or round in shape and have longitudinal reinforcing bars arranged in a circle and enclosed by a closely-spaced steel spiral. The ACI Code, which is the basis for most reinforced concrete design in this country, permits larger loads on spiral than on tied columns when loaded concentrically because of the greater toughness of spirally reinforced columns (correct answer A).

21. Match each symbol with the term which correctly describes it.

I. I/c 1. slenderness ratio
II. $1/r$ 2. modulus of elasticity
III. $\sqrt{I/a}$ 3. radius of gyration
IV. f/ε 4. section modulus

A. I-1, II-3, III-4, IV-2
B. I-2, II-1, III-3, IV-4
C. I-4, II-1, III-3, IV-2
D. I-4, II-2, III-1, IV-3

The exams frequently include questions on structural terms and symbols, and these questions are sometimes of the matching type, as in this case. I/c is called the section modulus (I-4), a term used in beam design. $1/r$ is the slenderness ratio (II-1), which is used in the design of columns. $\sqrt{I/a}$ is the radius of gyration (III-3), a property used in column design, and f/ε is the ratio of unit stress to unit strain, which is the modulus of elasticity of a material (IV-2). The symbols and the terms are correctly matched in answer C.

22. Most pile-driving formulas relate the capacity of a pile to all of the following, *except*

  A. the weight of the pile hammer
  B. the drop of the hammer
  C. the penetration of the pile per blow, under the last few blows of the hammer
  D. the angle of internal friction of the soil.

The bearing capacity of driven piles may be determined by a number of different methods, including the use of pile-driving formulas. All of these formulas base the capacity of a pile on the driving energy of the hammer: the product of the hammer weight and its drop (A and B) and the pile's penetration per blow (C). The angle of internal friction (D) is not a factor in pile-driving formulas; D is therefore the correct answer to this question.

23. In constructing a plywood stressed skin panel, the plywood is generally fastened to the lumber stringers with

  A. nails   C. wood screws
  B. glue   D. wood screws and glue

Flat panels with stressed plywood skins and spaced lumber stringers act like a series of built-up I-beams, with the plywood resisting most of the flexural stresses, while the lumber stringers act like the web and resist the shear stresses. The connection between the plywood and the lumber must be strong enough to resist the high shear stresses developed. While mechanical fasteners, such as nails or screws, might conceivably be used, they permit slip, which would result in excessive vertical deflection. For that reason, stressed skin panels must be fabricated with glue, which permits no slip to occur. B is the correct answer.

24. What is the portal method?

  A. A means of providing wind bracing in multistory buildings
  B. An approximate method for analyzing multistory building frames
  C. An approximate method for analyzing statically indeterminate beams
  D. A method for analyzing structures utilizing computer programs

The portal method is an approximate method for analyzing multistory building frames (correct answer B), in which several simplifying assumptions are made. These assumptions include the following:

1. *The horizontal shear is the same in all exterior columns.*
2. *The horizontal shear in each interior column is twice that in an exterior column.*
3. *The inflection points of all columns and beams are located midway between joints.*

In this age of the computer, with its tremendous speed and accuracy, isn't it archaic to use an approximate method of structural analysis? Not at all. Before a refined analysis of a multistory frame can be made, the member sizes must be estimated in order to know their relative stiffness. So as a starting point, some approximate analysis must be made. Also, it is often desirable to make a rough check of computer results, using approximate methods, to detect any gross computer errors.

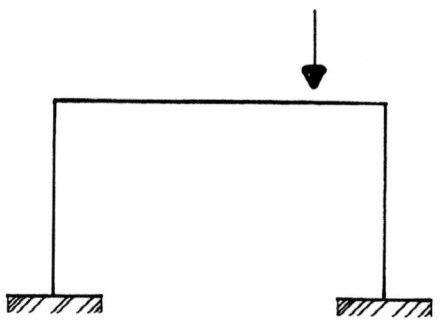

25. A one-story rigid frame is loaded as shown. Select the diagram which most closely approximates the shape of the deflected frame.

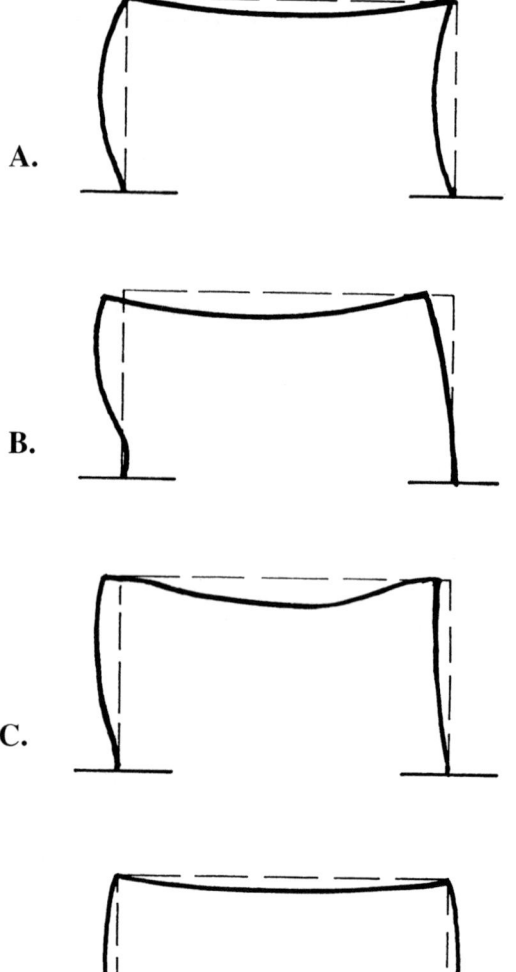

Having an intuitive grasp of how a structure deflects under load is a great asset in structural design. All is not lost, however, if one doesn't have that intuition—a few rules may suffice. For example, one should know that most rigid frames will move sideways when subject to either horizontal or vertical load, or both. Two exceptions: (1) when the structure is restrained from horizontal movement, and (2) when the frame is symmetrical and subject to a symmetrical vertical load. The vertical load in this question is not symmetrically placed, since it is to the right of the center line of the frame. Therefore, we should expect the frame to move sideways. That eliminates choices A & D. One should also know that when a column base is fixed, as in this question, a line tangent to the deflected column at its base will be vertical. This is true for B, but not for C, and so B is the correct answer.

VERTICAL TRANSPORTATION

1. Select the correct statement from among the following.

   A. Maximum elevator traffic in a building is generally computed at several different times during the day.

   B. Elevator traffic generally reaches its maximum peak during lunch time.

   C. Elevator capacity is computed on the basis of car size.

   D. Elevator capacity is computed on the basis of travel speed.

   This question could be deceptive if you are not entirely clear on the meaning of the terms used or how maximum traffic and capacity are determined. To begin with, both statements A and B are partially correct and deal with elevator traffic, which is computed on the basis of the number of passengers carried during a five minute span. Traffic is figured for peak or critical traffic periods, which vary considerably with the type and use of the building. For example, office buildings usually reach their maximum peak during lunch time (B), but hospitals might reach their maximum peak during visiting hours, and hotels reach theirs in the morning, early evening, and at check-out time (A). Elevator capacity, on the other hand, is based simply on car size (correct answer C) that is, the number of people an elevator car will comfortably hold. Travel speed (D) is considered only when computing elevator traffic-handling capacity, which requires both car size (the number of people carried) and round trip time, which is based on speed. In other words, capacity, which represents the number of people per car, and handling capacity, which represents the number of people carried during a five minute span, are two entirely different concepts.

2. In the design of vertical transportation, the most important factor is

   A. safety.

   B. speed.

   C. practicality.

   D. efficiency.

   Candidates should not have to spend much time pondering this question; because in this, as well as many other exam questions, the answer is based on the architect's responsibility to maintain the health, safety, and welfare of the public. Very simply, safety (correct answer A) comes first, and all other concerns are subordinate to that factor.

3. Elevators present special problems when fire breaks out in a multistoried building. In this situation, and with regard to safe emergency exiting, select the correct statement from among the following.

   A. In order for elevators to provide a safe means of exiting during a fire, they should always be enclosed in a fire-resistant shaft.

   B. Elevators should be designed so that cars will automatically return to the lowest floor, in the event of fire.

   C. Elevators should be designed so that they will travel in only the down direction, in the event of fire.

   D. Elevators should not be used for emergency exiting, in the event of fire.

   Elevators appear to be the fastest, most efficient way to evacuate a tall building in the event a fire breaks out. In most cases, however, elevators and fires don't mix. The problem is one of inadequate control and life safety. If, for example, passengers enter an elevator on a floor higher than one that is burning, there is a good chance the car will

129

be stopped, in its way down, by someone wanting to escape from a floor that is ablaze. In that event, passengers could be exposed to the fire danger. Therefore, elevators should never be considered for emergency fire exiting (correct answer D). Regardless of the structure's height, emergency fire stairways should be used instead. Incidentally, elevator shafts are generally fire-resistant, however, this does not imply that they are safe for emergency exiting when a fire breaks out.

4. Match the following apparatus with the appropriate range of speeds at which they normally operate.

   I. Hydraulic passenger elevator
   II. Electric passenger elevator
   III. Freight elevator
   IV. Escalator

   1. 50–200 fpm
   2. 90–120 fpm
   3. 50–150 fpm
   4. 300–1,800 fpm

   A. I-2, II-3, III-1, IV-4
   B. I-3, II-4, III-1, IV-2
   C. I-4, II-1, III-2, IV-3
   D. I-2, II-4, III-3, IV-1

This is certainly not an easy question, but it involves data for which a candidate should have some feeling. The analysis begins with the one range of speeds that differs radically from all the rest. That speed of up to 1,800 feet per minute (4) must be matched with the apparatus type that normally functions at the greatest speed—the electric passenger elevator (II). Since this match-up exists in two of the four answers (B and D), we can eliminate the other two choices A and C. The other type of equipment that one may know about is the escalator (IV), which operates at two standard speeds, 90 fpm and 120 fpm (2). This match-up is found only in the correct answer B. The other two pieces of equipment have similar operating speeds; however, one can assume that the hydraulic elevator is generally the slowest, since its overall travel distance rarely exceeds 50 feet. Some confusion may exist here, as we are not told whether the freight elevator is electric or hydraulic. Under the circumstances, we should assume it is electric, which implies that it has a greater speed of operation. Incidentally, these rates of speed are only approximate, and candidates may find variations from these figures.

5. Select the *incorrect* statement concerning escalators.

   A. Escalators convey passenger traffic more safely than do stairways.
   B. Escalators are capable of handling greater passenger traffic than elevators, in a given span of time.
   C. Although escalators are custom-built, they are standardized in width, angle of incline, and speed of operation.
   D. Since they are unidirectional, escalators are always installed in pairs, to accommodate both up and down traffic.

This question is almost a short lesson in escalator theory and design. First of all, escalators provide rapid, comfortable, and continuous vertical travel from one floor of a building to another. They are fast, efficient, and considerably safer than stairways (A), based on recent accident reports. Because they operate continuously, without the long waiting periods often associated with elevators, escalators can handle greater passenger traffic than elevators (B). Statement C is also true; escalators are fabricated in standard widths of 32 and 48 inches, their angle of incline is 30 degrees, and their speed of operation is either 90 or 120 feet per minute. The false statement and correct answer is D; escalators are usually, but not always, installed in pairs. There are some installations, for example, where traffic control is critical, in which escalators are provided in only one direction. Other installations provide escalators for up traffic and stairways for down traffic. Finally, there are instances where one escalator is provided and its travel direction changed from up to down, or vice versa, depending on the traffic flow. Examples of this type of installation are found in theaters and stadiums.

WOOD

1. Grading is the important process used to classify lumber. It is also true that grading

   A. is necessary because the properties of wood vary so considerably.

   B. is based on the strength, durability, and use of a particular species of wood.

   C. is based on criteria established at each mill by licensed graders.

   D. which was once primarily a visual process, is now accomplished largely by sophisticated instruments.

   The purpose of grading lumber is to establish reasonable standards of uniformity. These standards are necessary because wood, being a product of nature, varies considerably in its properties (correct answer A). In fact, no two pieces of wood have exactly the same strength, appearance, or physical characteristics. Although a small quantity of lumber is machine graded, grading is largely a visual evaluation process (D) based on criteria established by national lumber inspection bureaus, which are generally part of various lumber manufacturing associations (C). The grade of a piece of lumber is based on the number, location, and type of imperfections (B) that may adversely affect the wood's strength, durability, or use. In other words, grading seeks to identify defects, such as knots, splits, checks, and undesirable grain configurations. The best grades of lumber are nearly free from imperfections, while in each lower grade, the imperfections increase in quantity.

2. In conventional wood framed structures, the purpose of a purlin is to

   A. provide continuous support over openings cut in a wall surface.

   B. produce a positive connection between two opposite sloping rafters in the same plane.

   C. form the angle of a hip roof and support the jack rafter ends.

   D. support the roof rafters and, consequently, a portion of the roof load.

   Purlins are actually uniformly spaced beams or girders which are used to support roof rafters (correct answer D). They are used in wood framed roof construction in the same way girders are used in floor construction, to reduce the span of the regular members and help support the load. Purlins run at right angles to the rafters and may be used in either sloping or flat roof construction. The other statements in this question describe specific elements that are also used in wood framed structures. For example, statement A describes a header, statement B defines a collar beam or tie, and statement C refers to a hip rafter.

3. From among the qualities or properties listed below, select those that refer to plywood.

   I. Laminated and bonded.

   II. Any number of layers up to nine

   III. Grain of all plys parallel

   IV. Knotholes permitted

   V. Face and back veneers determine grade

   A. I, V   C. I, IV, V
   B. II, III   D. I, II, V

   Plywood is one of the most widely used forms of wood. This versatile material is laminated from thin wood sheets that are permanently bonded together with glue (I). The grain of the layers runs alternately at right angles (III), with the outside faces parallel to each other. Therefore, a plywood panel always requires an odd number of layers (II). All plywood is graded according to the quality of the face and back veneers (V), from A, which is the best to D, the poorest. Grades C and D permit small knotholes (IV), as this grade is generally used for sheathing, which is covered by other materials. The correct combination of qualities, therefore, is found in answer C.

4. The standard unit of measure for lumber is the board foot, which is defined as

   A. the volume of lumber in a 12" × 12" × 12" timber.
   B. one-twelfth of a cubic foot of lumber.
   C. a nominal one-inch thick board that is 12" wide.
   D. the weight of a 1" × 12" board that is 12" long.

A board foot of lumber measures quantity or volume, not weight (D), and it is used to measure, compute, and price lumber. It is defined as the amount of lumber contained in a piece of rough, green timber one inch thick, 12 inches wide, and one foot long. Correct answer B describes the same volume of lumber in a slightly different and possibly deceptive way. Incidentally, the board foot measure applies also to equivalent volumes of lumber which may be thicker, wider, narrower, or longer. For example, a 2" × 4" that is 18 inches long represents the same volume of wood as a 1" × 12" that is 12 inches long, which is exactly one board foot.

5. Select the incorrect statement concerning the moisture content of wood.

   A. In general, all woods shrink about the same amount, provided their moisture content is initially the same.
   B. In general, the moisture content affects the weight of wood.
   C. A reduction in the moisture content of wood increases its resistance to fungi attacks.
   D. A reduction in the moisture content of wood improves its structural capacity.

The moisture content of wood has a significant effect on several physical and mechanical properties of the material. To begin with, all lumber, when first cut from trees, has a high moisture content. Before it is used, it must be seasoned, which refers to the removal of moisture by air-drying or kiln-drying. Seasoning reduces the lumber's weight (B), increases its resistance to fungi and decay (C), and increases its strength (D). The only incorrect statement, therefore, is correct answer A. All wood does not shrink alike, regardless of its moisture content. In general, hardwoods shrink more than softwoods, and heavy woods shrink more than light woods, although there are rare exceptions to these general rules.